Posture Recognition for Fall Prevention
with
Low-Complexity UWB WBANs

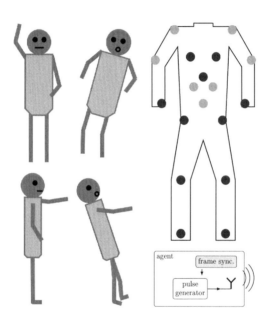

Robert Heyn

λογος

Series in Wireless Communications
edited by:
Prof. Dr. Armin Wittneben
Eidgenössische Technische Hochschule
Institut für Kommunikationstechnik
Sternwartstr. 7
CH-8092 Zürich

E-Mail: wittneben@nari.ee.ethz.ch
Url: http://www.nari.ee.ethz.ch/

Bibliographic information published by the Deutsche Nationalbibliothek

The Deutsche Nationalbibliothek lists this publication in the
Deutsche Nationalbibliografie; detailed bibliographic data are
available in the Internet at http://dnb.d-nb.de .

ISBN 978-3-8325-5621-1
ISSN 1611-2970

Logos Verlag Berlin GmbH
Georg-Knorr-Str. 4, Geb. 10, D-12681 Berlin
Tel.: +49 030 42 85 10 90
Fax: +49 030 42 85 10 92
http://www.logos-verlag.de

Abstract

This thesis concerns wearable ultra-wideband (UWB) wireless systems for human posture recognition. Our main objective is a technical proposal for a low-complexity realization of such a system. Posture recognition is a key enabler of various applications in the emerging field of technology-assisted approaches to maintain or improve human health. Especially fall prevention in the context of assisted living for elderly citizens relies on accurate body posture monitoring to detect and prevent imminent falls. Wireless body area networks (WBANs) pose a suitable means for continuous posture monitoring while preserving the user's privacy. Such wireless systems directly offer secondary uses such as sensor data transmission.

The difficulty of accurately modeling the time-varying wireless channels around the human body hinders analytical solution attempts to posture recognition. This thesis follows a measurement-based approach to posture recognition. For this purpose, we conduct a comprehensive measurement campaign. Therein we acquire the channel matrix between 18 body-mounted wireless nodes for a wide frequency range and a wide variety of postures. The postures are selected to cover an extensive range of daily activities. Measurements are performed in different environments for various test subjects of different physique in order to ensure a diverse and representative dataset. The acquired measurement data provides a solid foundation for a feasibility analysis and selection of key parameters of a posture recognition system for fall prevention from WBAN signals.

The constraints of a wearable battery-powered system require the on-body wireless hardware to be of low complexity. Based on the acquired high-resolution measurement data, we distinguish three different levels of measured raw data corresponding to different hardware complexity levels. In particular, we consider the complex-valued channel impulse responses, their magnitude, and the received signal energy for each on-body link. We evaluate suitable physical layer options to measure the respective information for each level, which is subsequently used for posture recognition.

In the course of a preliminary feasibility analysis, we develop a simplified system model and provide the corresponding maximum likelihood classifiers for the posture recognition problem for all three levels of measured data. Our analysis shows that posture classification based on the acquired data is feasible. It furthermore allows us to define an operating range of crucial parameters for the further system design.

For the classification of postures based on channel measurements, we consider a variety of standard machine learning algorithms. Based on their quantitative comparison, Random Forests are selected as a suitable classification method which can identify postures reliably from low-complexity energy measurements.

In order to further reduce the complexity and hardware demands, we minimize the number of required WBAN nodes. We demonstrate that accurate posture classification is possible with few suitably placed low-complexity nodes. This low-cost and low-complexity system design is thoroughly tested regarding its robustness towards additional disturbances and limitations of the data. The analysis includes an evaluation of the influence of the environment, cross-subject testing, and the reliability for the classification of unknown postures which are not present in the training data.

Based on the findings of this work, we propose a conceptual design for a wearable posture recognition system. We outline the requirements and limitations of this design proposal for fall prevention applications and provide a realistic performance assessment of its implementation.

Kurzfassung

Diese Arbeit befasst sich mit tragbaren drahtlosen UWB Systemen für die Erkennung der menschlichen Körperhaltung. Das Hauptziel ist ein Systemvorschlag für die Umsetzung eines solchen Systems mit geringer Komplexität. Die Erkennung von Körperhaltungen ist eine Schlüsselfunktion für verschiedene Anwendungen im wachsenden Bereich der technologie-assistierten Ansätze für Gesundheitserhaltung und -verbesserung. Insbesondere Sturzprävention im Kontext des betreuten Wohnens für ältere Menschen erfordert eine genaue Überwachung der Körperhaltung, um bevorstehende Stürze zu erkennnen und zu verhindern. WBANs stellen ein geeignetes Mittel zur kontinuierlichen Überwachung der Körperhaltung dar, welches die Privatsphäre der Nutzer schützt. Des Weiteren bieten solche drahtlosen Systeme direkten zusätzlichen Nutzen wie beispielsweise die Möglichkeit zur Übermittlung von Sensordaten.

Analytische Ansätze zur Erkennung der Körperhaltung werden dadurch erschwert, dass sich die zeitvarianten drahtlosen Kanäle um den menschlichen Körper herum kaum exakt modellieren lassen. Diese Arbeit verfolgt einen messbasierten Ansatz zur Haltungserkennung. Dazu führen wir eine umfassende Messkampagne durch, in der wir die Kanalmatrix zwischen 18 am Körper befestigten drahtlosen Knoten für einen breiten Frequenzbereich und eine Vielzahl von Körperhaltungen erfassen. Die Messungen werden in verschiedenen Umgebungen für verschiedene Testpersonen mit unterschiedlichem Körperbau durchgeführt, um einen diversen und repräsentativen Messdatensatz zu erhalten. Die erfassten Messdaten bilden eine solide Grundlage für eine Machbarkeitsanalyse sowie eine Auswahl der entscheidenden Parameter für ein Körperhaltungserkennungssystem zur Sturzprävention auf Basis von WBAN-Signalen.

Die Einschränkungen eines tragbaren, batteriebetriebenen Systems erfordern eine geringe Komplexität der am Körper getragenen Hardware. Ausgehend von den Messdaten in hoher Auflösung unterscheiden wir drei verschiedene Stufen von Rohdaten, die mit unterschiedlicher Hardwarekomplexität gemessen werden können. Im Detail unterscheiden wir die komplexwertigen Kanalimpulsantworten, ihren Betrag und die empfangene Signalenergie für jeden drahtlosen Kanal am Körper. Wir evaluieren geeignete Optionen für eine Physical-Layer-Implementierung, um die Rohdaten der jeweiligen

Komplexitätsstufe zu erfassen, die anschliessend für die Körperhaltungserkennung verwendet werden.

Im Verlauf einer ersten Machbarkeitsanalyse entwickeln wir eine vereinfachtes Systemmodell und geben die zugehörigen Maximum-Likelihood-Klassifizierungsmethoden für die Körperhaltungserkennung für alle drei Rohdatenabstufungen an. Unsere Analyse bestätigt die Machbarkeit der Körperhaltungserkennung auf Basis der gemessenen Daten. Des Weiteren ermöglicht sie, einen Betriebsbereich bezüglich der Schlüsselparameter für das weitere Systemdesign festzulegen.

Für das Klassifizierungsproblem der Zuordnung von Körperhaltungen zu den Kanalmessungen ziehen wir verschiedene Standardverfahren für maschinelles Lernen in Betracht. Basierend auf einem quantitativen Vergleich werden Random Forests als geeignete Klassifizierungsmethode ausgewählt, die die Körperhaltung basierend auf Energiemessungen mit geringer Komplexität erkennen kann.

Um die Komplexität und die Hardware-Anforderungen weiter zu reduzieren, minimieren wir die Anzahl der erforderlichen Netzwerkknoten. Es wird gezeigt, dass eine akkurate Haltungserkennung mit wenigen geeignet platzierten Knoten von geringer Komplexität möglich ist. Dieses günstige und wenig komplexe Systemdesign wird gründlich auf seine Robustheit gegenüber zusätzlichen Störeinflüssen und Einschränkungen der verfügbaren Daten getestet. Diese Analyse beinhaltet eine Betrachtung des Einflusses der Umgebung und einen Vergleich mit verschiedenen Testpersonen sowie eine Untersuchung der Zuverlässigkeit der Klassifizierung für Körperhaltungen, die nicht in den Trainingsdaten enthalten sind.

Basierend auf den Erkenntnissen dieser Arbeit wird ein konzeptionelles Design für ein tragbares System zur Erkennung der Körperhaltung vorgeschlagen. Wir stellen die Anforderungen und Einschränkungen des Systemvorschlags hinsichtlich der Sturzerkennung dar und geben eine realistische Abschätzung der zu erwartenden Leistungsfähigkeit einer entsprechenden Implementierung.

Acknowledgments

This work would not have been possible without the continuous support from my supervisor Prof. Dr. Armin Wittneben, whose guidance and support allowed me to grow both professionally and personally during the course of my doctorate. I am very grateful for the openness and flexibility in shaping the project, and for the opportunity to pursue my own research interests.

I would further like to express my immense gratitude to Prof. Dipl.-Ing. Dr. Klaus Witrisal for thoroughly reviewing this thesis as a co-referee and providing his valuable feedback.

I am further indebted to Dr. Marc Kuhn, who generously shared his knowledge and experience. His advice contributed significantly to the success of the project and offered new perspectives on arising challenges. My gratitude is extended to my colleagues Dr. Gregor Dumphart and Dr. Henry Schulten for their technical and personal support throughout the doctorate. Our informal but productive discussions in front of a whiteboard with a coffee often provided helpful insights and were very useful in shaping ideas and developing the next steps.

I am also grateful to Bharat Bhatia and Upasana Chakraborty for their support with the data acquisition, both for serving as test subjects and their assistance during the measurement campaign. Many thanks also to Oliver Stadler, Joël Küchler, and Roman Kunz, who conducted preliminary measurements in the laboratory environment under my guidance. The acquired data helped in gaining a better understanding of the challenges of the posture recognition task.

Finally, I would like to thank my family for the continuous support throughout these years. Especially the patience, understanding, advice and unconditional support from my dear wife Preethi has led this journey to its successful completion.

Zurich, September 2022

Contents

1

Introduction

1.1 Motivation

In ageing modern societies, an increasing number of elderly citizens is living independently. At a higher age, falls are a huge risk for the seniors' health and continued wellbeing. The growing manifold of technological approaches towards this issue offers a variety of solutions for supervision, allowing to detect when a fall has occurred. While fall detection potentially saves lives and reduces the response time to the fall incident, preventing the fall in the first place avoids the injuries and prevents long-term harm. Thus, fall prevention is clearly preferable over fall detection, albeit significantly more challenging. This is due to the fact that an imminent fall risk has to be detected within a sufficiently short period of time to initiate suitable countermeasures. Consequently, real-time monitoring of the body posture is required to identify fall-related postures in time to take preventive action.

Monitoring a user's body posture can either be achieved with posture estimation or posture recognition. Posture estimation, i.e. reconstructing an arbitrary posture from measurements, is very flexible but very challenging. Posture recognition on the other hand, i.e. the recognition of postures based on calibrated patterns, is significantly easier to implement. It further benefits from the recent advancements in the domain of machine learning but relies on and thus requires reliable training data. We will pursue the recognition approach in this thesis, after an earlier thesis [1] in the Wireless Communications Group at ETH Zurich explored the estimation approach.

A variety of sensors for posture monitoring is available, reaching from fixed radar- [2, 3] and camera-based systems [4–6] to wearable inertial sensors like accelerometers and/or gyroscopes [7]. Wireless signals have proven to be a suitable means for posture recognition both in the radio frequency domain [8] as well as in magneto-inductive systems [9–11]. Wearable systems of inertial or wireless sensors have the advantage over camera-based solutions of preserving the user's privacy. Furthermore, they are independent of external infrastructure and thus operate in all environments. Another important advantage of wireless signaling for posture recognition is the possibility for secondary use such as data transmission from other body-mounted sensors or radar applications to detect obstacles or monitor the gait [12–14]. The inherent communication capabilities enable the system to exchange data with a smartphone which provides computing resources and the possibility of integration into an app for the respective smartphone ecosystem.

Despite their advantages and the recent progress in the e-health domain, such assistance systems are still not as common as one might expect. A recent study on fall prevention and detection [15] identified user comfort as a key parameter for acceptance. It has further been found that low-cost and low-complexity systems are better received by the target group of elderly people [16]. This highlights the requirement for simple, affordable, and comfortable (wearable) fall prevention systems.

1.2 Related Work

A vast variety of literature is available in the context of fall prevention and posture recognition. Various surveys [17–19] structure and categorize the literature base in different ways, often with an individual focus such as low power systems [20]. In the following, we will provide a brief overview over the relevant works in relation to this thesis, particularly approaches which rely on radio frequency for posture recognition.

As outlined above, a diverse category consists of systems for ambient sensing without wearable components. Among radio systems, this includes UWB radar for posture recognition (e.g. [2, 3]) and radio signals between nodes in the surroundings [21, 22]. Mixed systems combine external infrastructure with body-worn nodes, e.g. using RFID [23]. Off-body channels for UWB systems have been modeled and analyzed e.g. in [24]. Aiming for ubiquitous operation without external infrastructure, we require a fully wearable system relying solely on on-body channels. UWB WBAN channels for four links between the wrist and other body parts have been analyzed by Abbasi et al. [25]. They consider different arm postures and compare different environments, namely an anechoic chamber and an indoor scenario. Their focus lies on the channel model, propagation effects and different modulation schemes for communication. Similar analysis was done by others, e.g. in [26, 27]. In a more recent work, a system proposed by Huang et al. [28] recognizes postures from UWB-based distance measurements between 14 body-mounted receivers and a ceiling-mounted transmitter array. While the system is versatile regarding the postures, it requires the fixed transmitters and a large number of nodes to reconstruct the body posture accurately.

Einsmann et al. [29] have simulatively explored the feasibility of a posture capturing system based on on-body time-of-flight measurements. Their study identifies potential for such a system to reconstruct arbitrary body postures and examines suitable sensor locations. Their approach is limited to a purely simulative posture estimation study, however.

Farella et al. [30] developed a wireless posture recognition system based on accelerometers. In their practical implementation [31] they provide detailed information of the WBAN to merge sensor information, including hardware and a power consumption analysis. The combination of an accelerometer-based system with posture recognition based on the signaling itself is a very attractive option to increase reliability and robustness.

Quwaider and Biswas [32] combine accelerometers with on-body narrowband received signal strength indicator (RSSI) measurements to also recognize low-activity postures, which are difficult to identify solely using accelerometers. They coarsely classify postures into four groups (stand, sit, walk, run). For tests with three subjects, they report high accuracy and outperform a threshold-based approach.

With an application of firefighter operations in mind, Geng et al. [33] aim to identify particular activities such as crawling or climbing a ladder from narrowband on-body signals. Various statistical characteristics of the on-body channels in the time and frequency domain serve as features for the classifying support vector machines (SVMs). Experiments in different indoor locations provide the basis for a detailed analysis of the importance of the features and the on-body links. In addition, the authors compare their approach to different accelerometer-based methods. Identifying static postures is challenging for both the proposed approach as well as accelerometer-based systems with few features, but less important for the intended field of application, i.e. firefighter operations. For posture monitoring in the context of healthcare applications, however, static postures are equally important. Furthermore, identifying critical fall-related postures requires a posture recognition from a snapshot measurement and does not leave sufficient time for observing statistical channel characteristics during a longer period.

Paschalidis et al. [34] have developed a posture recognition approach based on Generalized Likelihood Tests, which is compared to an SVM for classification based on narrowband on-body signals. Due to the fluctuations of the measured RSSI of the three links considered, a larger number of measurements is required for reliable classification. Although the authors provide little information on the measurements, it becomes clear that the focus lies on distinguishing similar standing postures. An analysis of the topology or possibilities of wideband measurements are not performed, and the measurement campaign appears limited despite the emphasis on a measurement-based approach.

In the Wireless Communications Group at ETH Zurich, recent publications have successfully demonstrated posture recognition for a limited set of postures using purely passive magneto-inductive coils [9–11]. With individual advantages (e.g. material penetration) and disadvantages (e.g. strong sensitivity towards metallic surroundings), a combination with conventional radio (as presented in this work) has potential to increase robustness of the posture recognition.

The most closely related work is by Yang et al. [35], who demonstrate human posture recognition based on on-body signaling in an anechoic chamber as well as an office environment with accuracies of about 80 % and 75 %, respectively. Their selection of postures, however, is substantially smaller than in this work, and their system only uses narrowband measurements in the 2.4 GHz-band, leaving the possibilities and the analysis of wideband solutions out of scope. Furthermore, their WBAN is limited to six links between the hip and various body parts. A systematic analysis of suitable node placement is not performed. Overall, their work has a variety of insightful initial approaches to various aspects of the topic, such as different environments or hyper-parameter tuning. However, their analysis remains superficial. We thus acknowledge it as a starting point and provide a more comprehensive, generalized, and detailed analysis of the topic of wearable wireless posture recognition in this thesis, enriched by additional aspects on e.g. the implementation and the robustness of such a wearable system.

Identification of shortcomings:

To summarize the state of the art, the existing literature provides a variety of approaches towards recognizing body postures from wireless signals. Wireless channels around the human body have been studied for different environments and postures. However, we have identified several gaps in the existing work: Posture recognition essentially relies on comprehensive calibration data to identify postures of various daily activities. The existing studies mostly lack a comprehensive and diverse set of training data, and/or are limited to few groups of postures. This does not allow for a feasibility evaluation of concrete problem setups like fall prevention for elderly citizens. Furthermore, very few approaches explore the benefits of wideband signaling. Especially with ultra-wideband capabilities of modern smartphones, it is worthwhile to explore possibilities beyond the 2.4 GHz-band. A systematic topology analysis or reduction is lacking, despite the importance of a low number of on-body sensors for comfort and cost of the system. Furthermore, most presented systems are not analyzed regarding their robustness and limitations, reporting high accuracies for a single test person and/or environment without examining a core challenge of posture recognition, i.e. the extension towards unknown postures.

1.3 Contributions

In this work, we propose a design for a wearable posture recognition system based on WBAN signals between on-body nodes. In the course of our measurement-based approach, we conduct an extensive measurement campaign to obtain the necessary data for this task. For a variety of 43 postures from various daily activities including fall-related postures, we measure the complete wideband channel matrices between 18 body-mounted nodes. We obtain a diverse dataset by recording data for all postures performed by three subjects of different physique in varying indoor environments.

We conduct a preliminary feasibility analysis based on the acquired data, for which we develop a simplified system model. For different levels of measured raw data, we provide the corresponding maximum likelihood classifiers for the system model. In addition to a demonstration of the feasibility of posture recognition based on WBAN signaling, we use the simplified system model for an evaluation of a suitable parameter space regarding signal-to-noise ratio (SNR) and frequency range.

From a variety of established machine learning approaches, we identify suitable classifiers for the posture classification task based on a quantitative comparison. For our favored selection (Random Forest classifiers based on UWB energy measurements), we conduct a deeper analysis of the parameter space for operation. This includes a systematic minimization of the WBAN topology, i.e. the number and placement of on-body nodes. Furthermore, we evaluate the robustness of the proposed configuration towards test data from different subjects, environments, and previously unknown postures which are not present in the training data. Finally, we propose a feasible concept for a wearable posture recognition system implementation based on our findings, and provide a simulative estimate of its expected performance.

1.4 Publications

In the course of this work, the following papers have been published: A pre-study with measurements in an anechoic chamber [8], a detailed introduction to our large-scale measurement campaign [36] (see Chapter 4), and an analysis of the feature importance for energy-based classification [37] (see Chapter 7).

- [8] R. Heyn and A. Wittneben, "Detection of Fall-Related Body Postures from WBAN Signals," in *GLOBECOM 2020 - 2020 IEEE Global Communications Conference*, 2020, pp. 1–6

- [36] R. Heyn and A. Wittneben, "Comprehensive Measurement-Based Evaluation of Posture Detection from Ultra Low Power UWB Signals," in *IEEE International Symposium on Personal, Indoor and Mobile Radio Communications (PIMRC)*, 2021, pp. 1–6.

- [37] R. Heyn and A. Wittneben, "WBAN Node Topologies for Reliable Posture Detection from On-Body UWB RSS Measurements," in *ICC 2022 - 2022 IEEE International Conference on Communications (ICC)*, 2022, pp. 1–6

Outside the context of this thesis, the following papers have been published: A study on user tracking with Bluetooth Low Energy (BLE) [38] (as first author), an analysis of antenna influences and diversity for BLE localization [39] (as co-author), and the proposal and implementation of a novel approach for distance and displacement estimation from UWB measurements [40] (as co-author) as an extension of its initial proposal in [41].

- [38] R. Heyn, M. Kuhn, H. Schulten, G. Dumphart, J. Zwyssig, F. Trösch, and A. Wittneben, "User Tracking for Access Control with Bluetooth Low Energy," in *2019 IEEE 89th Vehicular Technology Conference (VTC2019-Spring)*, 2019, pp. 1–7

- [39] H. Schulten, M. Kuhn, R. Heyn, G. Dumphart, F. Trösch, and A. Wittneben, "On the Crucial Impact of Antennas and Diversity on BLE RSSI-Based Indoor Localization," in *2019 IEEE 89th Vehicular Technology Conference (VTC2019-Spring)*, 2019, pp. 1–6

- [40] G. Dumphart, R. Kramer, R. Heyn, M. Kuhn, and A. Wittneben, "Pairwise Node Localization From Differences in Their UWB Channels to Observer Nodes," *IEEE Transactions on Signal Processing*, vol. 70, pp. 1576–1592, 2022

2

Posture Selection for Activities of Daily Living and Fall Situations

In this chapter, we introduce the relevant postures for the application of fall prevention for elderly citizens. We explain the differences and similarities of the selected postures and categorize them in a systematic way.

2.1 Significance of Calibration Hypotheses

A reasoned selection of postures is crucial for the posture recognition studied in this thesis. *Recognition* or *classification* means finding the best match for a sample of test data in a training dataset. This matching is based on certain features in the data, which are compared between the test sample and the training data according to certain metrics.

In our posture *classification* case, we compare characteristics (features) of the measured wireless on-body channels for a single posture (test data) to previously measured posture hypotheses (training data). The set of hypotheses is also referred to as "calibration data" in the following, as it is recorded during a calibration phase under known circumstances and with certainty about the posture. Posture *estimation*, on the contrary, is independent of previously obtained information and aims to reconstruct any arbitrary body posture based on measured information. While posture *estimation* is thus more versatile, it typically requires a more complex system to infer distances, angles etc. from measurements. Posture *classification* reduces the problem to the recognition of measured posture-dependent patterns in the hypotheses.

Although it is always possible to find the best match for a test measurement in the calibration data, it cannot be not guaranteed that the test data stem from the same posture. If a posture under test is not included in the hypothesis selection, the classification is bound to fail. Consequently, it is essential for posture recognition systems to include all expected test postures in the calibrated hypotheses in order to ensure reliable and correct classification.

For fall prevention approaches, it is thus required to anticipate all fall-critical postures of subjects in order to build a sufficient calibration dataset. This does not only imply all possible directions towards which a person could lose balance, but also major differences in their limb orientation and position: A measurement of a posture with stretched arms may differ significantly from a measurement of the otherwise same body posture with bent or folded arms.

We introduce a hierarchical order to our posture selection. The basic separation is by *category*, i.e. we assign all postures to either "activities of daily living (ADL)" or "fall-related". Within each category, we distinguish various *group*s of postures: ADL postures are separated into the groups "standing", "walking", and "sitting", where the latter includes the transitions of standing up and sitting down. Fall-related postures are assigned a group according to the direction of the imminent fall, which can be "forward", "backward", or "sideways". As explained in more detail in Chapter 4, the process of recording data for a single posture with the selected equipment requires several seconds, rendering it impossible to record postures continuously during a dynamic motion such as walking. Consequently, we split dynamic activities into multiple static snapshots.

A particular posture may occur during various activities: An elderly person could sit either at a table and eat or in an armchair and read, where the sitting posture is mostly identical but the environment differs. We account for these variations with a modification of the environment, such as the placement of a table. We refer to such a constellation and orientation of surrounding items at specific distances as a "setup" in the following. Measurements for every posture are conducted in various setups in order to take different situations and locations into account in which the posture might occur. Default setups for every posture are an open setup without additional items in immediate vicinity, i.e. closer than 1 m, and a cluttered setup with various items on at least two sides of the subject. Some postures require additional setups in order to model the circumstances of particular activities. These additional setups are described along with the respective postures in the following.

2.2 Activities of Daily Living

The first category comprises postures of common ADL of elderly residents. The postures are selected such that most common daily tasks can be assigned at least one of the postures or a variation of it. Note that sports activities are not included in our selection.

2.2.1 Standing Postures

The group of standing postures is the largest and most diverse group within the ADL category. Falls are more likely to happen when the patient is standing or walking [42], which renders a distinction of the respective postures from fall-related situations crucial. Our selection comprises ten different postures, which are shown in Fig. 2.1.

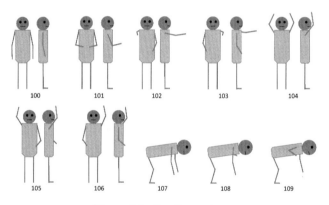

Figure 2.1.: Standing postures

The first posture is a relaxed upright standing posture with arms hanging down. Typical occurrences of this posture include situations of waiting, orienting oneself, and the moment after standing up from a seat. For the latter scenario, we include two additional setups, one with a chair behind and one with a chair behind and a table in front of the subject. The process of standing up/sitting down is covered in Section 2.2.3. A very similar posture with bent arms additionally covers the scenario of kitchen work or standing at a counter. We conduct measurements in three additional setups: With a chair behind and a table in front, with a metal chair in front (simulating a walking aid) and in front of a table counter. This is also a typical posture of a person holding items while standing. We further include two postures with each arm stretched forward, which occurs e.g. when reaching out to take an item from a shelf. Three postures with one or both arms elevated, respectively, represent a variety of situations, such as eating, talking on phone, and reaching items stored at a height. Lastly, the process of bending down in order to pick up items from the ground is represented by three postures: Reaching down with the right arm, the left arm, and both simultaneously. The ten standing postures with their respective setups are summarized in Table 2.1.

posture	description	example situation	additional setups
100	upright, arms down	waiting	chair behind, chair behind / table in front
101	upright, arms bent	kitchen work	chair behind / table in front, metal chair in front, table in front
102	right arm extended forward	taking item from shelf	
103	left arm extended forward	taking item from shelf	
104	both arms up	getting dressed	
105	right arm up	phone call	
106	left arm up	touching face	
107	bending, both arms down	picking item from ground	
108	bending, right arm down	picking item from ground	
109	bending, left arm down	picking item from ground	

Table 2.1.: Overview over standing postures (no. 1xx)

2.2.2 Walking Postures

The second group of ADL postures includes the ten walking snapshots shown in Fig. 2.2. We split a single step into three posture snapshots, resulting in six postures for a full walking cycle of two steps. Identifying the phase of a step, i.e. the particular leg position, can be important for fall prevention as it allows to infer the patient's weight distribution, which can help to anticipate the fall direction and/or adapt the respective countermeasures such as a muscle stimulus. In addition to walking on even ground, we consider postures for climbing up and down stairs, each with either foot forward, resulting in four additional postures. Table 2.2 summarizes the walking postures.

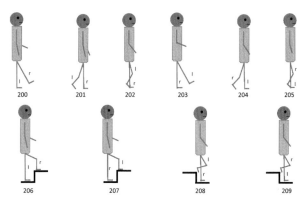

Figure 2.2.: Walking postures

posture	description
200	right foot elevated
201	right foot placed forward
202	feet parallel, weight on right foot
203	left foot elevated
204	left foot placed forward
205	feet parallel, weight on left foot
206	step up right
207	step up left
208	step down right
209	step down left

Table 2.2.: Overview over falling postures (no. 2xx)

2.2.3 Sitting Postures

Although the risk of an immediate fall is reduced while sitting, it is important to include sitting postures in the calibration data in order to avoid false positives in the detection of fall-related postures. We distinguish seven different sitting postures, which are shown in Fig. 2.3. A relaxed posture with the hands resting in the persons lap while leaning back typically occurs during watching television. While reading or eating, the posture varies: The upper body is more upright and the arms bent while holding an item. Measurements for this posture are taken in an additional setup with a table in front of the person. The same applies to the postures with the right and left arm close to the head, respectively, which are postures occurring while eating or making phone calls. Further, we include three snapshots of standing up/sitting down in the group of sitting postures: Standing up forwards from a free-standing chair and standing up forwards/sidewards towards right or left while taking support from a table. The last sitting posture is bending down forward with arms extended downwards, which is a typical posture performed e.g. when wearing shoes. Table 2.3 summarizes our selection of sitting postures.

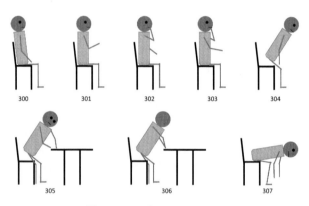

Figure 2.3.: Sitting postures

posture	description
300	leaning back, arms in lap
301	upright, arms bent
302	upright, right arm up
303	upright, left arm up
304	sitting down, arms down
305	standing up towards right, arms on table
306	standing up towards left, arms on table
307	leaning forward, arms down

Table 2.3.: Overview over sitting postures (no. 3xx)

2.3 Fall-Related Postures

We distinguish the fall-related postures by the direction of the imminent fall (forward/backward/sideways). Knowing the fall direction enables suitable countermeasures, such as triggering a selective muscle stimulus to prevent the fall or launching an appropriately placed wearable airbag.

2.3.1 Falling Forward

Tripping over small objects or steps is likely to result in a forward fall. Our selection of forward-falling postures mainly differ in the orientation of the arms, which are directed sideways, forward, upwards, or backwards, as shown in Fig. 2.4. Leaning towards one side, e.g. during a turn, may result in a fall towards a forward-left or forward-right direction, which is also covered in our posture selection. Finally, we include a posture during a fall forward, i.e. *after* the subject has lost balance, which will serve for verification purposes. We consider this posture beyond a point where a body-mounted system can prevent a fall entirely. However, measures to reduce the impact of the fall such as raising arms to protect the head may still be taken at this point. In order to maintain a constant posture during the measurement, we take support from a non-conducting rope which is fixed to the environment.

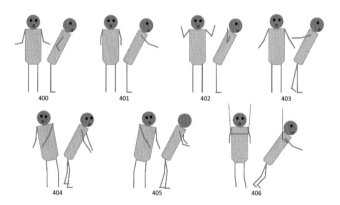

Figure 2.4.: Forward falling postures

2.3.2 Falling Backward

An incorrect shift of body weight is another common cause for falls [42], which can also happen in a backward direction. Our selection contains six backward falling postures, which are shown in Fig. 2.5. The distinct difference again is the arm position (forward/sideways/backwards), while the knees are slightly bent for all of the postures. We further include postures related to falling diagonally towards left and right backwards, as well as an actual fall for verification as described for the forward falling case. A fixed strap again provides support to maintain a constant posture during the measurement.

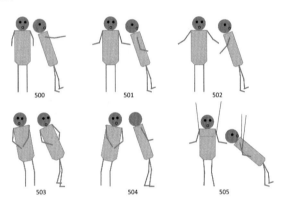

Figure 2.5.: Backward falling postures

2.3.3 Falling Sideways

Lastly, we include two postures related to sideways falls, which are shown in Fig. 2.6. Falls towards forward-left, forward-right as well as backward-left and backward-right have been included in the forward- and backward fall-related postures, respectively. Hence only the direct fall towards either side is included in the "sideways" group.

2.4 Mapping Activities to Selected Postures

Most common ADL of elderly people are directly included in our selection. In order to illustrate this, we will briefly examine a default daily routine of activities, indicating

600   601

Figure 2.6.: Sideways falling postures

group	posture	description
Forward	400	arms downwards
	401	arms stretched forward
	402	arms raised
	403	arms backwards
	404	arms forward/down, towards right
	405	arms forward/down, towards left
	406	arms forward, lost balance
Backward	500	arms forward
	501	arms down/sideways
	502	arms back
	503	arms down, towards right
	504	arms down, towards left
	505	arms sideways, lost balance
Sideways	600	towards right
	601	towards left

Table 2.4.: Overview over falling postures (no. 4xx, 5xx, 6xx)

the corresponding postures in parentheses.

After waking up, the patient sits on the bed (300, 301), stands up (304) and walks to the bathroom (200-205) for the morning hygiene, including changing clothes (100-106), freshening up (100-106) and using the toilet (304, 300). Preparing food (100-103) and sitting down at a table to eat (304, 301-303, 305) as well as leisure activities like watching TV (300), reading (301) and making phone calls (302, 303) can be mapped to the respective postures, too. In addition, picking things from the floor (107-109), stowing items away (102-106), even getting ready to go out by changing clothes (102-106) and wearing shoes (306) are covered with the proposed selection. The evening routine including bathroom and sitting on/in bed repeats the postures of the morning routine in different order. As mentioned above, we exclude unusual activities like sports in this thesis.

2.5   Summary

This chapter has introduced the postures of our datasets in a structured manner. We distinguish between two posture categories (ADL and fall-related), which are further divided into groups depending on the main activity and the direction of the imminent fall, respectively. In order to account for different situations in which postures may occur, measurements need to be conducted in various setups, i.e. with different constellations of surrounding items.

3

Physical Layer: Raw Data Availability & System Complexity

This chapter briefly introduces the core concept for our envisioned posture recognition WBAN. We further discuss different levels of WBAN channel state information as raw data and their availability at the on-body receiver nodes, outlining the trade-offs regarding the hardware implementation complexity. Existing physical layer approaches are briefly summarized and evaluated regarding their suitability for the posture recognition WBAN at hand. Moreover, we specify timing requirements and further assumptions regarding the parameters for our evaluation.

3.1 Network Concept

3.1.1 Hierarchical Structure

The posture recognition task at hand requires multiple wireless nodes distributed across the user's body. In order to optimize for a low node complexity and low power consumption, we distinguish different roles of the wireless nodes. Nodes with a limited functionality are tailored for a dedicated role and can hence be stripped of all components not related to their core functionality to reduce the power consumption. We divide the nodes into three groups with distinct functionality, which are hierarchically ordered by their required complexity. The hierarchical concept for a WBAN for posture capturing has been discussed in the context of this work in [8], and is summarized in the following. Chapter 8 covers some further details regarding the implementation which are omitted in this conceptual introduction.

The three complexity levels are in the following referred to as "agents", "anchors", and "central unit". The first level with the lowest complexity consists of agent nodes broadcasting a wireless signal. Agents are characterized by low power consumption and low cost, making them affordable for use in larger numbers. The broadcasted agent signals are received by the second level nodes, i.e. the anchors. We refer to this transmission as the "first hop". Due to their higher complexity, anchors can not only receive but optionally aggregate or pre-process signals before forwarding them to the third level, i.e. the central unit. This forwarding transmission is referred to as "second hop" in the following. Agents and anchors have to be body-mounted, as the characteristics of the first hop transmissions are used for posture recognition and thus need to be posture-dependent and reproducible. The central unit can be located off-body and does the computationally more demanding processing, especially the actual posture classification. Hence, a smartphone poses an attractive choice due to its computational resources and possible integration of other health monitoring functions.

Implanting the body-mounted nodes subcutaneously is attractive from the system designer's point of view in order to avoid exposure to mechanical impact or variations of the node position on the body. Charging the batteries requires wireless power transfer to the subcutaneous in-/on-body nodes, which has been discussed e.g. in [43, 44]. A more easily acceptable approach from the user's point of view is the integration of the system in clothing and/or wearables. Integration of electronics into garments conveniently ensures wearability and even allows for wired connections between nodes on the same piece of fabric (e.g. for time and frequency synchronization), but requires the respective clothing to be worn in order for the system to operate. This renders it impossible to prevent falls in certain situations (e.g. in the bathroom).

The node placement on the body for both a hierarchical (3-level) and a transceiver-based (2-level, with nodes combining the agent and anchor functionality) architecture is analyzed in Section 7.3 in Chapter 7.

3.2 Data Processing Approaches

The posture recognition problem at hand can be approached in various ways. In the following, we aim to provide a brief overview and categorization. Fig. 3.1 illustrates the different techniques and schematically explains their respective processing workflow. Potential pitfalls are highlighted in red.

We distinguish between *Statistical Modeling* and *Machine Learning* approaches. The former refers to methods based on the evaluation of probability density functions

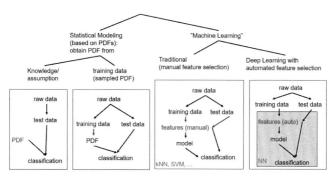

Figure 3.1.: Data processing approaches

(PDFs) to find the maximum likelihood solution for the classification problem. Consequently, accurate information about the PDFs is crucial. The PDF is either found by assumption (e.g. Gaussian), Monte Carlo simulation, or obtained from samples of measured training data. Each of these has potential pitfalls (highlighted in red in Fig. 3.1): The assumption of a particular PDF may be incorrect or oversimplified. Simulations may not be sufficiently accurate to provide a valid basis for the classification of real-world measurement with unforeseen disturbances, which are not accurately modeled. If the PDF is obtained from measured training data, a large amount of data is required to ensure a correct representation of the PDF. In the WBAN-based posture recognition context, this approach has been evaluated in [34].

Machine learning approaches have become increasingly popular in recent years. While they can be classified according to a manifold of different criteria, we simply distinguish traditional machine learning approaches with manual feature selection from deep learning approaches using neural networks, which include the feature selection in their processing. The common principle is the classification of test data based on features, i.e. characteristics or patterns which have previously been observed in training data. The feature space, i.e. the selection of features used for classification, determines the separability of the classes (postures in our case). Manual feature selection is often based on data availability, intuition, and/or trial and error. The reduction of available data to a limited set of features comes with an information loss, which may pose a drawback to the classification if valuable information is discarded. The training data set comprising labeled (i.e. assigned to a true class/posture) samples of features is used to train a model ("supervised learning"), i.e. find appropriate decision boundaries between the classes in the feature space. Based on their position in the feature space, unlabeled test data samples are then assigned to the respective class. An introduction to machine learning including common classifiers like SVMs, k-nearest neighbors (kNN) etc. is provided e.g. in [45]. Posture recognition based on wireless on-body signaling with machine learning approaches has been explored e.g. in [33,35]. Chapter 6 includes an introduction in the respective terminology and methods.

In deep learning, neural networks use the entire available data from the training set to find the most suitable feature space. The automated selection makes it possible to find patterns in the data which are not obvious to the human eye. However, it requires careful parameter tuning and a large amount of training data, and the networks' black box characteristic makes an analysis and debugging of the process challenging. Nevertheless, deep learning become popular for a wide range of classification tasks [46]. We evaluate one deep learning approach for the posture classification task in Section 6.4.2 in Chapter 6.

3.3 Level of Information from Available Raw Data

Posture recognition in this work is based on a comparison of a test measurement with reference measurements, also known as training data. Consequently, the information captured in the measurement process plays a crucial role for the posture recognition. In turn, the obtainable information depends on the capabilities of the measurement device. In our case, the measured variables are the channel impulse responses (CIRs) between on-body nodes.

We distinguish three levels of information to be captured from the CIR in the measurement process: The complex-valued CIR, its magnitude, and the energy aggregated over a certain time window. These levels are selected with a focus on the hardware required to measure the respective information. The trade-offs between the amount of measurable information and the complexity of the measurement hardware are explained later in this chapter.

Available information on each level is fully contained in the following levels. The conversion functions f_i to obtain each level i from the complex-valued CIR $h(t)$,

$$f_1(t) = h(t) \tag{3.1}$$

$$f_2(t) = |f_1(t)|^2 = |h(t)|^2 \tag{3.2}$$

$$f_3(t_0, t_1) = \int\limits_{t_0}^{t_1} f_2(t) \, dt = \int\limits_{t_0}^{t_1} |h(t)|^2 \, dt, \qquad (3.3)$$

are not bijective, which implies that there is information loss in the transition from one level to a later one. In turn, the required measurement hardware at the receiver side can be simplified for raw data levels with less information. In the following, we will provide a brief overview of the requirements for measuring the respective raw data. The following introduction is intended as a reminder for the reader, without delving into the details of specific implementations.

3.3.1 Level 1: Complex-valued CIR

Coherent receivers capture both amplitude and phase information, i.e. the complex-valued CIR. Fig. 3.2 shows a block diagram representation of a coherent receiver for channel estimation with a Nyquist pulse. The received and bandpass-filtered signal $x(t)$ is downconverted into baseband in the inphase and quadrature branch. The matched filter output of each branch is sampled with the symbol rate in the analog-to-digital converter (ADC). Note that timing recovery for the sampling instances could be performed before and after the downconversion [47]. A brief overview and a derivation of the maximum likelihood estimations for phase and timing are provided in chapter 5 of [48], further details on a variety of receiver structures can be found in [49]. A key parameter characterizing the obtainable information about the CIR as well as the complexity of the hardware implementation is the bandwidth of the signal. Interference from other users in the same frequency range can be expected especially in the crowded unlicensed frequency bands. For this work, we assume any frequency offset as constant for the duration of the CIR, resulting in a phase offset of the signal. Synchro-

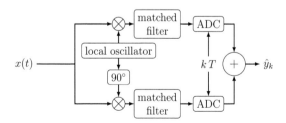

Figure 3.2.: Block diagram of a coherent receiver

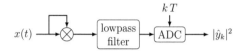

Figure 3.3.: Block diagram of a noncoherent receiver

Figure 3.4.: Block diagram of an energy receiver

nization errors in the phase and timing recovery circuits can deteriorate the receiver performance, but a detailed analysis is beyond the scope of this work. The reader is referred to [47] for an extensive analysis. The assumptions for this work regarding timing synchronization are explained in Section 3.5.

3.3.2 Level 2: Magnitude of the CIR

The receiver structure is simplified for the noncoherent case, as shown in Fig. 3.3. Analogously to the complex-valued CIRs, the signal bandwidth and the resulting Nyquist sampling frequency determine the time resolution and the complexity of the implementation. This affects especially the sampling ADC.

3.3.3 Level 3: Signal Energy

Measuring only the signal energy $\int_{t_0}^{t_0+T_{\mathrm{win}}} |h(t)|^2 \, dt$ in a fixed time window $[t_0, t_0 + T_{\mathrm{win}}]$ requires the lowest complexity. Ignoring the phase information and the fine timing information contained in the signal omits phase and timing recovery, as shown in Fig. 3.4. Consequently, only a coarse timing synchronization is required to turn on the receiver for the time of an expected signal reception. This is especially attractive for wireless sensor networks with battery-operated nodes, whose reduced duty cycle allows to save energy between transmissions and thus extend the battery life. While stripping the receiver of complexity, the information obtained from a measurement is very limited compared to the CIR sampled at the Nyquist rate. Characteristics such as the number and delays of multipath components (MPCs) are not available. Its simplicity makes this type of receiver very attractive for our application. Sampling at the Nyquist rate is not required, and the choice of the sampling rate is independent of the signal bandwidth. Shorter sampling intervals increase the time resolution of the instantaneous energy profile. If the sampling rate of the energy measurement is shorter than the delay spread of the channel, the energy distribution over the different time bins even provides coarse information about the MPCs.

3.4 Physical Layer Implementation Candidates

3.4.1 Narrowband Candidate: Bluetooth Low Energy

As an example for narrowband candidates, we summarize BLE in the following. The IEEE specifies a variety of other physical layer implementation options for WBANs in their 802.15.6 standard [50].

The current Bluetooth standard, which is maintained by the Bluetooth Special Interest Group, is version 5.2 [51] and was published in 2019. For low power short-range communication, BLE has been introduced in version 4.0 of the standard. A characteristic of BLE is the advertising mode using three channels of $1\,\text{MHz}$ bandwidth each, allowing to operate in a broadcasting mode with the receiving counterpart ("scanner") listening on these advertising channels. Thus the Bluetooth-typical pairing process is skipped and data is transmitted directly. Instead of maintaining multiple piconets between the on-body nodes, we consider the advertising mode of BLE for the link measurements.

3.4.1.1 Data Availability

The standard specifies a minimum receiver sampling interval of $1 - 2\,\mu s$, the transmission of an uncoded BLE packet takes $44 - 2128\,\mu s$. Advertising events are repeated every $\approx 20 - 30\,\text{ms}$, which poses a tolerable limit on the update rate of the on-body channel measurements in our envisioned system as the reaction time for human reflexes, i.e. "natural fall prevention", lies in the same range.

Multipath propagation in indoor environments induces small-scale fading of BLE signals [39], and the small bandwidth of the advertising channels does not permit a separation of the different signal propagation paths. As a result, the measurement of level 1 information results in a single complex channel coefficient for a time interval of $\frac{1}{B} \approx 1\,\text{ms}$. Additionally, BLE records the RSSI upon every packet reception.

With the Constant Tone Extension, the BLE standard further specifies the option to include simple beamforming for Angle-of-Arrival (AoA) and Angle-of-Departure (AoD) for devices with multiple antennas. IQ-demodulation enables phase recovery for this purpose [51].

3.4.1.2   Multiple Access

The 40 available BLE channels are used for frequency divison multiple access with frequency hopping to reduce interference [51]. Further, additional time division via polling is described in the standard. A large number of advertising agent nodes increases the risk of collisions in the three primary advertising channels. Using the remaining 37 data channels as (secondary) advertising channels mitigates this shortage of frequencies and would allow to support additional agents.

3.4.2   Wideband Candidates

3.4.2.1   Direct Sequence Spread Spectrum

Direct Sequence Spread Spectrum (DSSS) is an established approach e.g. in the third generation of mobile networks, the second WiFi generation, and military applications [52]. By multiplying each baseband symbol with a pseudo-random noise (PN) sequence of N_{chip} chips $c_k = \pm 1$ so that $T_{\text{symb}} = N_{\text{chip}} T_{\text{chip}}$, a signal is spread over a bandwidth of $1/T_{\text{chip}}$. Spreading is typically done in the baseband, but is also possible after upconversion to a carrier [48]. Linear feedback shift registers generate the PN sequences for spreading.

Correlating the downconverted received signal with the same (synchronized) PN sequence despreads the signal and restores the original baseband symbols as $c_k \cdot c_k = 1$ for all chips c_k. The processing gain from the correlation is $G \approx \frac{T_{\text{chip}}}{T_{\text{symb}}} = N_{\text{chip}}$. The correlation of the received signal with the reference sequence at the receiver can be performed in either the analog or the digital domain. Analog correlation requires very precise timing synchronization. Digitalizing the sequences requires sampling at least at the chip rate ($T_{\text{samp}} \leq T_{\text{chip}}$).

The spreaded and 2-PSK-modulated signal $s(t)$ to transmit N_{symb} symbols using the base bulse $g(t)$ can be written as

$$s(t) = \sum_{i=1}^{N_{\text{symb}}} \sum_{k=1}^{N_{\text{chip}}} b_i c_k g(t - kT_{\text{chip}} - iT_{\text{symb}}) \cos(\omega t) \tag{3.4}$$

where $b_i, c_k \in \{\pm 1\}$ determine the polarity of the base pulse $g(t)$ for every symbol and chip, respectively.

3.4.2.2 Data Availability

The received signal is filtered with the base pulse and the spread training sequence and sampled at the chip rate to obtain a channel estimate. Filtering the received signal with time-shifted copies of the PN sequence (typically shifted by integer multiples of T_{chip} [48]) can be parallelized. Noncoherent receivers benefit from rapid acquisition techniques such as the approach presented in [53], which reduces the need for many parallel operations at the receiver and is especially efficient for Gold codes [54, 55].

The chip duration is a crucial parameter for both the required sampling rate (as $T_{\text{samp}} \leq T_{\text{chip}}$) and the clock speed for the sequence-generating shift register. The signal bandwidth of the spread signal $B \approx 1/T_{\text{chip}}$ also scales inversely to the chip duration [48]. As the transmission of a PN sequence requires $T_{\text{seq}} = N_{\text{chip}} T_{\text{chip}}$, a short chip duration allows for either (i) a shorter transmission time and thus a lower duty cycle, or (ii) longer sequences (and thus a higher correlation gain) for the same transmission time. However, there is a tradeoff with the increasing hardware complexity for the resulting larger bandwidth and faster switching/sampling.

3.4.2.3 Multiple Access

Code Division Multiple Access (CDMA) is inherent in DSSS systems. Every transmitting node uses an individual PN sequence as unique identifier (UID), which is used to despread the signal at the receiver, thus allowing all nodes to use the same frequency range simultaneously. Without further division in time or frequency, a received signal needs to be correlated with all UIDs in order to find the respective matching agent node which transmitted the received signal. Interference between users is avoided by selecting PN sequences with low cross-correlation as agent UIDs.

3.4.2.4   Ultra-wideband Impulse Radio

UWB has gained popularity for a wide range of applications, such as indoor localization [40,56], radar [57] and WBANs [58]. Signals with a bandwidth of at least 500 MHz or 20 % of the center frequency (fractional bandwidth) are referred to as UWB signals by the FCC [59]. The frequency range between 3.1 and 10.6 GHz has been assigned for UWB usage, with strict regulations of the radiated power to ensure interference-free coexistence with other (conventional) systems.

Transmitting simultaneously over such a wide frequency range provides a number of advantages: The large bandwidth allows for high data rates and increases the robustness towards narrowband interference. Precise ranging via time-of-flight measurements and resolving MPCs becomes possible with the very short pulses. UWB nodes can be designed for an extremely low power consumption due to the low duty cycle. A reference design for an ultra low power receiver is given in [60]. These advantages and the recent integration in smartphones [61] makes UWB an attractive candidate for the posture recognition application. A more extensive introduction to UWB is provided e.g. in [62,63]. The IEEE standard 802.15.4 [64] defines the respective UWB channels for different pulse repetition rates and modes of operation.

3.4.2.5   Data Availability

However, UWB transmission poses challenges to the receiver: Coherent reception of the extremely short pulses requires timing synchronization in the sub-nanosecond range and high quality analog components, which are typically costly and power-consuming [65]. Rake receivers utilizing the variety of MPCs are difficult to implement in the analog domain. Digital processing, however, requires very high sampling rates according to the Nyquist theorem: If the full UWB bandwidth of 7.5 GHz is used, the signal must be sampled with at least 15 $\frac{\text{GS}}{\text{s}}$. Consequently, coherent UWB receivers require a very high complexity.

With our focus on low complexity solutions for WBANs, UWB impulse radio (IR) is of particular interest as it allows for simple noncoherent receiver architectures [66]. The simplicity however comes at the cost of performance compared to complex coherent approaches [67]. IR baseband pulses are transmitted directly without upconversion or modulation onto a carrier [68]. Hence, the transmitter only requires a component for pulse generation, such as a step-recovery diode or avalanche transistor, and a pulse shaping filter. The shaped baseband pulse is directly fed to the transmit antenna [63]. The trade-off for the simple architecture is a suboptimal spectral efficiency.

On the receiver side, an energy detector (ED) receiver can measure the signal energy with very low complexity [65,66]. The received signal is filtered, squared and integrated over a fixed time window, which is typically longer than the delay spread of the channel. Consequently, the energy contributions from all relevant MPCs are included. The sampled output of the integrator is stored as a channel characteristic (for the posture recognition) or compared to a threshold (for bit detection in communication).

As mentioned in Section 3.3.3, a coarse approximation of the energy of the CIR over time is possible by separating the integration window into shorter bins (shorter than the delay spread of the channel) [69]: For instance, sampling the energy detector output at $1\,\frac{GS}{s}$, the receiver can obtain the energy profile over time, averaged over 1 ns per sampling instance. MPCs with a smaller delay difference cannot be separated, but a rough approximation for CIRs with a delay spread longer than the bin duration (i.e. sampling period) can be obtained. Timing estimation based on the sampled ED output was demonstrated in [65].

3.4.2.6 Multiple Access

A common multiple access approach for noncoherent UWB IR receivers is time hopping (TH) [70], which additionally helps to smoothen distinct lines in the signal spectrum [67]. The individual pulses are shifted in time according to a user-dependent hopping pattern within a fixed time interval (pulse frame). The ED receiver adapts its integration window to select the respective user, which requires sufficient time synchronisation between the transmitters and the receiver.

An alternative approach is Direct-Sequence UWB (DS-UWB) [62], which works in the same fashion as DSSS using bipolar UWB pulses for chips. As the ED receiver cannot determine the polarity of a received pulse, an autocorrelation receiver is required, which can be seen as a (noncoherent) extension of the ED receiver [67].

3.5 Timing

The Vector Network Analyzer (VNA) measurements described in Chapter 4 are fully synchronous in time, phase and frequency due to the measurement capabilities of the VNA. In a fully wireless on-body system, we can expect asynchronicity especially regarding the time reference between the nodes. Hence, in a wireless system of lower complexity a link measurement

$$\hat{h}_{ij}(t) = h_{ij}(t - \Delta\tau_{ij}). \tag{3.5}$$

of the CIR $h_{ij}(t)$ between nodes i and j includes the link-individual timing offset $\Delta\tau_{ij}$. The offset $\Delta\tau_{ij}$ includes both nodes' individual clock offsets and is hence correlated to other links to/from nodes i and j. Timing offsets from imperfect synchronization and other signal processing delays from both nodes shall also be included in τ_{ij}.

The literature provides a wide variety of synchronization techniques to estimate this offset [47,71,72], with trade-offs between performance and complexity. Our proposed agent–anchor–central unit constellation differs from typical point-to-point communication systems. Low complexity synchronization for wireless sensor networks is studied in [73], and for relaying systems in [74]. An UWB IR synchronization scheme for ED receivers is presented in [75].

As an extensive analysis of timing and synchronization is beyond the scope of this work, we use the following timing noise model for our evaluations: We assume time synchronization which is sufficiently precise for the detectable part of the CIR to always be within the observation window. Any unknown offsets between measurements are modeled as random cyclic shifts of the sampled CIR vector. These random shifts are generated individually for every node. They are independent and identically distributed according to a uniform distribution over an interval τ_{shift}, which is a parameter depending on the assumed timing synchronization between the nodes. For every link, the sum of the individual node offsets is discretized according to the time resolution of the VNA and applied as a cyclic shift to the CIR vector.

3.6 Summary

In this chapter, we have outlined the trade-offs between data availability and implementation complexity for measuring CIRs at receiver nodes. Important aspects of our WBAN application have been discussed and promising physical layer candidates have been introduced. In the remainder of this thesis, we will focus on UWB as a physical layer due to its aforementioned advantages and its underrepresentation in the literature on posture recognition. The complexity of a particular implementation depends on the choice of several design parameters. In the following, we indicate their ranges of interest for our application.

Frequency Range

The signal bandwidth determines the time resolution of the measured CIR and hence the separability of multipath components. A large bandwidth further makes the system more robust against narrowband interference. The center frequency determines the average path loss and possible interference from other users in the same frequency band (e.g. the 2.4 GHz-ISM band). The antenna size scales with the signal wavelength $\lambda_c = \frac{c_0}{f_c}$. For the posture classification datasets, we consider the frequency range of $3.1 - 8.5$ GHz.

Timing Offset

The timing synchronicity between the nodes is modeled by a time shift of the measured CIRs. Depending on the synchronization technique, these offsets can be estimated and corrected to a certain extent. We assume a sufficiently fine synchronization between nodes to guarantee that the CIR lies within the observation window.

Measurement Duration

The measurement duration should ideally be as long as the delay of the last relevant multipath component. A shorter time window does not capture all relevant characteristics of the CIR, whereas a longer window collects additional noise. This is especially crucial for the low complexity energy measurements. We vary the observation time between 40 ns and 1 μs.

Energy Sampling Frequency

Measuring the CIRs coherently (level 1) or their magnitude (level 2) requires sampling at the Nyquist rate. For the energy measurement, the sampling frequency can be reduced to relax the requirement for the ADC. In this work, we use a single measurement of each link energy, i.e. the sampling frequency equals the snapshot measurement rate.

4

Measurements

This chapter characterizes the hardware used for data acquisition and provides a detailed insight into the measurement procedures of the project. We further provide details on the measurement environment and the test subjects, and discuss some observations from the acquired data.

Relation of S-parameters and the Channel Impulse Response

As a brief reminder for the reader, we revisit the relationship between Z-parameters, scattering parameters (S-parameters) and the frequency response of a system in the following. A more detailed introduction can be found e.g. in [76].

We consider a point-to-point transmission link from a source to a load over a wireless channel. The physical representation of the system is illustrated in Fig. 4.1, consisting of a source with internal impedance Z_{src} feeding a transmitting antenna, a wireless channel (indicated by the arcs), and a load Z_{load} connected to a receiving antenna. The relationship between the input voltage v_1 at the transmitter and the observed voltage v_2 at the receiver is described by

$$\underbrace{\begin{bmatrix} v_1 \\ v_2 \end{bmatrix}}_{\mathbf{v}} = \mathbf{Z} \underbrace{\begin{bmatrix} i_1 \\ i_2 \end{bmatrix}}_{\mathbf{i}} \tag{4.1}$$

with the Z-matrix $\mathbf{Z}$ of the two-port representation of the system as shown in Fig. 4.1b.

The voltages v_i at each port i can be described as the sum of an incoming voltage wave v_i^+ and an outgoing voltage wave v_i^-, which are defined with respect to a reference impedance Z_0 as

$$\mathbf{v}^+ = \frac{1}{2}\left(\mathbf{v} + Z_0\,\mathbf{i}\right) \qquad \mathbf{v}^- = \frac{1}{2}\left(\mathbf{v} - Z_0\,\mathbf{i}\right), \tag{4.2}$$

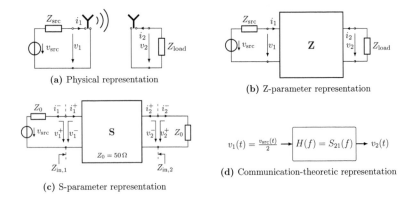

(a) Physical representation

(b) Z-parameter representation

(c) S-parameter representation

(d) Communication-theoretic representation

Figure 4.1.: System model for communication link

so that $\mathbf{v} = \mathbf{v}^+ + \mathbf{v}^-$ and $\mathbf{i} = \mathbf{i}^+ - \mathbf{i}^-$, where $\mathbf{i}^+ = \mathbf{v}^+ Z_0$ and $\mathbf{i}^- = \mathbf{v}^- Z_0$. The scattering matrix of S-parameters describes the relation of incoming and outgoing voltage waves at all ports as

$$\mathbf{v}^- = \mathbf{S}\,\mathbf{v}^+. \tag{4.3}$$

Consequently, the S-parameters depend on the choice of Z_0, which is typically selected as the characteristic impedance of the connecting cables ($Z_0 = 50\,\Omega$). Fig. 4.1c shows the equivalent representation for this case. As $v_2^+ = 0$ (i.e. no incoming wave at port 2) in our case with $Z_{\mathrm{load}} = Z_0$, we only have an outgoing voltage wave v_2^- regardless of the input impedance of port 2. The incoming wave at port 1 does not depend on the input port impedance, either: For a source impedance $Z_{\mathrm{src}} = Z_0 = 50\,\Omega$, we have with (4.2) and $v_1 = v_{\mathrm{src}} - Z_{\mathrm{src}}\,i_1$

$$v_1^+ = \frac{1}{2}\left(v_1 + Z_0\,i_1\right) = \frac{1}{2}\left(v_{\mathrm{src}} - Z_0\,i_1 + Z_0\,i_1\right) \overset{Z_{\mathrm{src}}=Z_0}{=} \frac{v_{\mathrm{src}}}{2}. \tag{4.4}$$

Note that a mismatch between $Z_{\mathrm{src}} = Z_0$ and the input impedance of port 1 $Z_{\mathrm{in},1}$ leads to an outgoing (reflected) wave v_1^-. This is observed in the scattering parameter S_{11} (commonly referred to as "reflection coefficient"), which is unequal to zero in case of a mismatch between Z_{src} and $Z_{\mathrm{in},1}$. With the previous observation that $v_2 = v_2^-$ as

$v_2^+ = 0$ and (4.4), we have

$$v_2 = v_2^- = S_{21} \cdot v_1^+ = S_{21} \cdot \frac{v_{\text{src}}}{2}, \tag{4.5}$$

which allows us to link the S-parameter description of our system to the well-known communication-theoretic signal model as shown in Fig. 4.1d, assuming $Z_{\text{src}} = Z_{\text{load}} = Z_0 = 50\,\Omega$. We refer to the parameter S_{ji} with $i \neq j$ interchangeably as "transmission coefficient" or "frequency response", and to its squared magnitude as "path loss" or "channel gain" for a wireless scenario with antennas at both input and output port.

The extension of the two-port example in Fig. 4.1 to a multiport is straightforward. For the measurement scenario with 18 on-body antennas, the equivalent description is an 18-port. With all equipment (antennas, cables, and VNA) matched to $50\,\Omega$, this applies to the measurement setup in this work. Hence, the measurement of $S_{21}(f)$ over a range of frequencies is equivalent to the frequency response $H(f)$. The corresponding CIR $h(t)$ is obtained using an inverse Fourier transform. Note that the impedance matching is worse for frequencies below 3 GHz, as can be seen from the higher antenna reflection coefficient in Fig. 4.5 in the following section.

4.1 Antennas

All measurements are conducted with identical planar UWB antennas printed on a 1 mm thick FR4 substrate of $35 \times 28\,\text{mm}^2$ based on the design presented in [77]. Fig. 4.2 shows the front and back side of the printed circuit board. The antenna is matched to $50\,\Omega$ and equipped with an SMA connector to connect to a coaxial cable. For a center frequency of $f = 5\,\text{GHz}$, the wavelength is $\lambda_0 = c_0/f = 6\,\text{cm}$ in free space and $\lambda_r = \frac{c_0}{\sqrt{\varepsilon_r} f} \approx 3\,\text{cm}$ in the substrate with a relative permittivity of $\varepsilon_r \approx 4$. The Fraunhofer distance marking the beginning of the far-field [78, p.10] is thus $d_f = \frac{2D^2}{\lambda_0} \approx 4\,\text{cm}$ with the largest antenna dimension $D = 35\,\text{mm}$.

4.1.1 Influence of Feeding Cables

As we aim to emulate a network of wireless nodes, the influence of the required antenna feed cables potentially poses a nuisance. The radiation from cables attached to printed antennas is a known issue [79] and can be prevented with adapted antenna design including slots or a hidden feed line [80]. As the antennas available for our measurements

 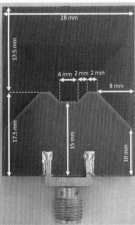

(a) Front (antenna)　　　　　(b) Back (ground plane)

Figure 4.2.: Planar UWB antenna

do not feature such designs, we take a different approach to reduce the influence of the cables close to the antenna connectors by covering the respective cable sections with foamy polyurethane absorber [81].

We quantify the influence of the leakage current radiation on the feed cables for our antennas with the following experiment: For two vertically oriented antennas positioned in parallel on tripods at a distance of 0.5 m, we measure the frequency response for different arrangements of the feeding cables: Vertical parallel feeding cables from below, and horizontally fixed cables which are orthogonal to the antennas, feeding them from the opposite site of the respective other antenna. For both cable orientations, we measure the frequency response with and without absorber foam around the cables. The relative antenna orientation is kept constant.

Fig. 4.3a shows the path loss differences $\Delta|S_{21}(f)|^2$ between the open and the absorber-clad feeding for both setups (vertical and horizontal feed). We distinguish the case of *open* cables ($|S_{21}(f)|^2_{dB,horz,open} - |S_{21}(f)|^2_{dB,vert,open}$), *absorber-clad* cables ($|S_{21}(f)|^2_{dB,horz,abs.} - |S_{21}(f)|^2_{dB,vert,abs.}$), and *horizontal* cables with and without absorbers ($|S_{21}(f)|^2_{dB,horz,open} - |S_{21}(f)|^2_{dB,horz,abs.}$). The results are averaged over 10 measurements each and smoothened by time-windowing the CIRs to 20 ns.

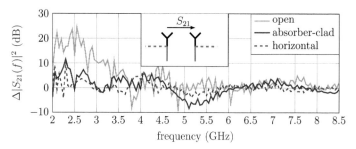

(a) Channel gain for vertical antennas with different feed cable orientation

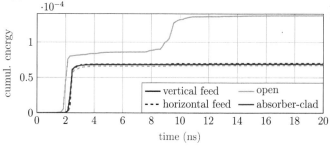

(b) Cumulated energy of the CIR for different feed cable orientations

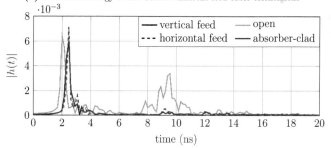

(c) CIRs for different feed cable orientations with absorber

Figure 4.3.: Measurements for open and absorber-clad cables

We observe significant differences between the setups without absorber foam, especially below 3.5 GHz. With absorber foam around the feeding cables, the difference is reduced by multiple decibels. A similar difference in the measured channel gain is observed when comparing horizontal feed cables with and without absorbers (dashed). This is expected, as the (dashed) cables are aligned on the same axis as indicated in the illustration in Fig. 4.3a. Hence, radial radiation from leakage currents is not received at the other antenna. Hence, we conclude that the absorber foam minimizes undesired radiation from the leakage currents. This is confirmed by the strong similarity of the cumulated energy $(E_{\text{cumul.}}(t) = \int_0^t |h(t)|^2 dt)$ profiles of the CIRs in Fig. 4.3b, and the CIRs in Fig. 4.3c.

Consequently, in order to mitigate the effects of feeding cables on the measured frequency responses, the feeding cables can either be attached orthogonally to the antenna, or the radiating portion of the feeding cable can be cladded with absorber. Orthogonal cable management while mounting the antennas at fixed positions and parallel to the body surface (like antennas integrated in clothing) was found particularly challenging especially on the torso. Hence, we follow the latter approach and shield the cables at the antenna feed point.

In order to determine a suitable absorber length, we perform the following experiment: We measure the channel gain for two vertical antennas mounted on a tripod at 0.5 m distance, with vertical feeding cables. Again, the fixed setup on the tripod allows to reproduce the exact positioning and eliminates influence of the human body, so that the effect of the absorbers can be studied separately. We repeat the channel gain measurement for absorber cladding of 15 cm and 30 cm around the parallel feeding cables. The frequency responses are averaged over 10 measurements each and smoothed by time-windowing the CIRs to 20 ns for clarity of illustration. Fig. 4.4 shows the path loss difference $\Delta |S_{21}(f)|^2$ between 15 cm and 30 cm absorber cladding. We observe that the measured channel gain does not change significantly with additional absorber.

We thus conclude that the majority of radiation from leakage currents occurs close to the connector, and that absorber cladding of 15 cm around each cable is sufficient to mitigate this undesired radiation. For all following measurements, the feeding cable sections closest to the antennas are shielded with 15 cm of absorber foam. Feeding cables are attached parallel to the antennas for convenience of mechanical fixation, especially on the torso.

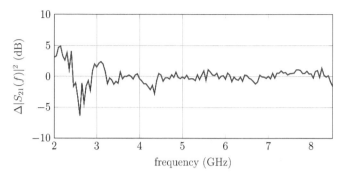

Figure 4.4.: Channel gain difference between 15 cm and 30 cm absorber cladding around feeding cables

4.1.2 Reflection Coefficient

Fig. 4.5 shows the antenna's frequency-dependent reflection coefficient $|S_{11}|^2$, averaged over 10 measurements and smoothened with a time window of 20 ns. A reflection coefficient of $|S_{11}|^2 \leq -10\,\text{dB}$ is often considered suitable for antenna operation [82–84]. For $f > 3.1\,\text{GHz}$, the reflection coefficient is mostly below the threshold of $-10\,\text{dB}$. We further observe that the immediate surroundings of the antenna only have a minor influence on the reflection coefficient.

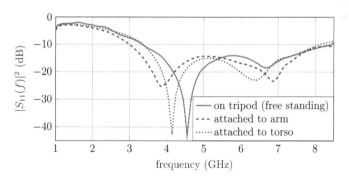

Figure 4.5.: Reflection coefficient $|S_{11}|^2$

4.1.3   Radiation Pattern

For the azimuth and elevation planes as defined in Fig. 4.6, the direction-dependent antenna gain is shown for four frequencies in Fig. 4.7 in comparison to an isotropic radiator (dBi). Despite the planar structure, the radiation pattern is that of a typ-

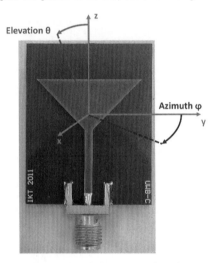

Figure 4.6.: Definition of azimuth and elevation angle with antenna in the y-z-plane

ical dipole, with an approximately rotationally symmetric azimuth pattern and deep notches in the elevation pattern. Fig. 4.7 further illustrates the frequency dependency of the radiation patterns, with variations of several decibels across the UWB frequency band and angles.

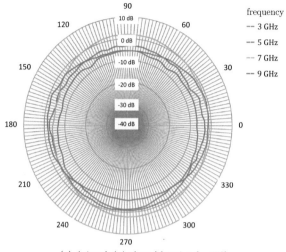

(a) Azimuth (φ) plane (elevation $\theta = 90°$)

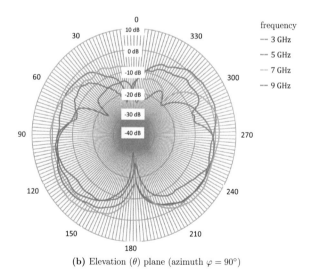

(b) Elevation (θ) plane (azimuth $\varphi = 90°$)

Figure 4.7.: Radiation pattern of UWB antenna [85]

4.1.4 Path Loss

The expected transition between near- and farfield is at a free-space distance $d \approx 4\,\text{cm}$ as shown earlier. In order to characterize the propagation environment around the human body, we briefly examine the distance dependence of the path loss $|S_{21}|^2$ in the farfield. According to Friis' equation, we have

$$\frac{P_{\text{RX}}}{P_{\text{TX}}} = G_{\text{TX}}G_{\text{RX}}\left(\frac{\lambda_0}{4\pi d}\right)^2, \tag{4.6}$$

where we use $G_{\text{TX}} = G_{\text{RX}} = G$ from Fig. 4.7 at $\varphi = \theta = 90°$. We measure the path loss at various distances for both free-standing tripod-mounted antennas as well as between antennas attached to the inside of the wrists. In all cases a line of sight (LOS) is maintained, and the antennas are aligned parallel vertically. Fig. 4.8 shows the path loss for 3 GHz and 7 GHz with the indicated antenna gain of $-5\,\text{dBi}$ and $1\,\text{dBi}$, respectively, as shown in Fig. 4.7. The results are averaged over 10 measurements each. We observe a good agreement between measurements and the expected free

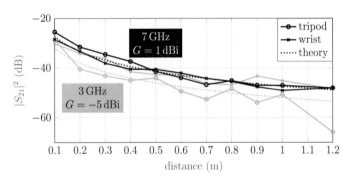

Figure 4.8.: Distance-dependent path loss $|S_{21}|^2$

space propagation. Note that this cannot be expected to apply to non-line of sight (NLOS) links, which occur between various body-mounted antennas. This is covered later in this chapter.

4.1.5   Polarization

A polarization mismatch between transmit and receive antenna can significantly deteriorate the channel gain. Hence, it is worthwhile to consider the polarization sensitivity of the antennas. For this purpose, we measure the path loss $|S_{21}(f)|^2$ between two identical antennas mounted in parallel on a tripod at a distance of 0.5 m. Fig. 4.9 illustrates the channel gain for different relative antenna orientations: "Parallel", i.e. both antennas oriented vertically, and "orthogonal", i.e. one antenna oriented vertically and the other antenna horizontally. The data is averaged over 10 measurements each and smoothened by time-windowing the CIR to 20 ns. The antennas are placed opposite to each other at respective azimuth and elevation angles of $\varphi = \theta = 90°$, so that the observed difference in the channel gain occurs due to the polarization. With

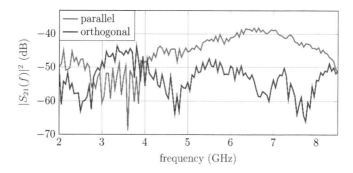

Figure 4.9.: Frequency response for different antenna orientations

a maximum difference of about 20 dB between parallel and orthogonal orientation, the influence of the linear polarization is apparent. It is worth pointing out that for our application, the linearly polarized antennas are suboptimal: A minor change in the posture can induce a rotation of the polarization, e.g. by a turn of the wrist. As a consequence, the amplitude of links to this particular node can vary significantly, which impacts the fingerprint of the respective posture. We incorporate the impact of the polarization for our posture measurements by constantly varying every posture within reasonable boundaries across several measurements, including different angles of the extremities where appropriate. This way multiple relevant relative antenna orientations are considered for every posture, potentially creating a more robust fingerprint.

4.1.6 Placement

The postures introduced in Chapter 2 are complex body constellations which differ from each other to varying degrees. In order to reliably distinguish especially similar postures which differ only in minor details, i.e. the position and orientation of few body parts, it is crucial to ensure that the collected data contains information about these details. Consequently, antennas must be placed all over the body to capture variations of the position of every body part, resulting in a large number of antennas. We refer to the wireless connection between two antennas as a "link" in the following, and use the terms "antenna" and "node" interchangeably. The total number of links N_{link}, i.e. all possible antenna pairs, depends on the number of nodes N_{node} and is determinded as

$$N_{\text{link}} = \frac{1}{2} \sum_{n=1}^{N_{\text{node}}} (N_{\text{node}} - 1) = \frac{N_{\text{node}}(N_{\text{node}} - 1)}{2}. \tag{4.7}$$

The factor $\frac{1}{2}$ is due to to the reciprocity of the links, i.e. the complex frequency response is independent of the direction of the wireless transmission: $S_{ij}(f) = Sji(f)$. Consequently, measuring a link in a single direction is sufficient.

For our measurement campaign, we distribute $N_{\text{node}} = 18$ antennas across the test subject's body, resulting in $N_{\text{link}} = 153$ inter-node links. Fig. 4.10 shows the node placement schematically, which follows a number of considerations outlined in the following.

4.1.6.1 Density

It is crucial to include all posture-determining parts of the body in the measurements, which requires antennas on all segments of the flexibly moving limbs and around the torso. For instance, consider the walking postures (2xx), some of which differ only in the position of a single limb element (e.g. the lower leg segment in postures 200 and 202), requiring antennas at the ankles for distinction. Similarly, the position and orientation of the arms demands antennas attached to wrists and elbows to distinguish e.g. the different standing postures (1xx), whose leg position is mostly identical. The nodes on the torso serve as fixed points to determine the limb orientation relative to the torso, as required e.g. to distinguish postures 102 and 105 (standing with arm forward and upward, respectively). Some redundancy between the nodes on the chest and between

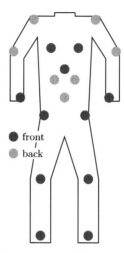

Figure 4.10.: Node placement on the body with nodes on front (dark) and back (light)

the nodes on the back is expected. In order to determine suitable placement of the torso-mounted antennas, we use all six nodes on the torso during data acquisition and reduce the topology at a later stage (cf. Chapter 7). A total of 18 nodes is sufficient to cover all relevant body parts. Furthermore, the distribution ensures several LOS connections, which is beneficial for stable channel conditions. NLOS links are expected to show more variation across multiple measurements of the same posture.

4.1.6.2 Intrusiveness & Comfort

A wearable system must not be a hindrance to daily activities or cause discomfort to the user in order to be accepted and worn constantly. Thus, node placement should be as unobtrusive as possible. Although there is scope for smaller antenna designs and thus smaller wireless nodes, we follow the idea of an unobtrusive placement and fix the antennas slightly off the joints. We further omit the head for comfort, as our postures are majorly determined by the relative positions of torso and limbs.

4.1.6.3 Reproducibility

The placement next to the limb joints comes with the additional advantage of more static antenna positions in our wired setup throughout the test subject's posture or movements. Variation of the antenna positions is reduced to a minimum, which ensures a more realistic emulation of antennas integrated in tight clothing or implanted subcutaneously.

4.2   Vector Network Analyzer

While oscilloscopes are commonly used for time domain signal analysis, VNAs conduct coherent measurements in the frequency domain. One VNA port at a time serves as an output and is fed a stimulus signal while the complex-valued frequency response at all other ports serving as inputs is observed. The frequency of the stimulus is swept over a user-defined range. All ports of interest successively take the role of the stimulated output port in order to obtain the full (frequency-dependent) network matrix.

In Section 4.1 we pointed out the requirement of the antenna distribution across all relevant body parts, which results in a large number of links. Measuring multiple links simultaneously comes with several advantages, as outlined in Section 4.3, but requires a VNA with as many ports as on-body antennas to avoid any manual switching or reconnecting ports to antennas during a measurement cycle.

In this work we use a Rohde & Schwarz ZNBT8 multiport VNA [86] to measure the S-parameters between the antennas. The supported frequency range for the 24 ports is 100 kHz to 8.5 GHz. We apply the configuration in Table 4.1 for all our measurements.

parameter	symbol	value
minimum frequency	f_{min}	2 GHz
maximum frequency	f_{max}	8.5 GHz
resulting sweep bandwidth	B	6.5 GHz
number of frequency sweep points	N_{freq}	6501
resulting frequency step	Δf	1 MHz
resolution/intermediate frequency (IF) bandwidth	RBW, B_{IF}	1 MHz
output power	P_{out}	0 dBm
nominal port impedance	$Z_{in,VNA}$	50 Ω
reference impedance for s-parameters	Z_0	50 Ω

Table 4.1.: VNA configuration

The VNA is calibrated with the set of coaxial antenna cables using the Rohde & Schwarz ZN-Z154 calibration unit [87] so that any effect of the cables on the transmission such as attenuation and phase shift is compensated. The reference plane is thus moved to the connection point of coaxial cable and antenna. Consequently, the antennas are considered a part of the measured channel.

4.2.1 Noise Characteristic

In the following, we will briefly characterize the measurement noise of the VNA, and relate the VNA measurement results to the SNR performance which we expect from a UWB system.

The magnitude of the autocorrelation function of the measured VNA noise is shown for the time domain in Fig. 4.11a and for the frequency domain in Fig. 4.11b. The noise measurement is conducted with a tripod-mounted antenna while the stimulus port of the VNA is terminated with a $50\,\Omega$-impedance. The peak of the autocorrelation

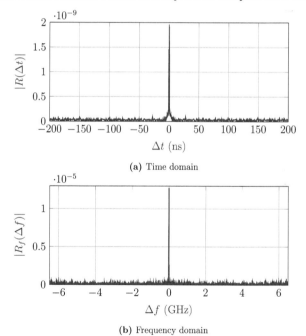

(a) Time domain

(b) Frequency domain

Figure 4.11.: Autocorrelation functions of VNA noise

function (with $|R_t(\Delta t)| \geq 4 \cdot 10^{-10}$) in time domain is $\approx 0.9\,\text{ns}$ wide. The peak in the frequency domain comprises a single frequency bin. The consistently low values of the autocorrelation functions for larger time and frequency shifts confirm the independence assumption of the measurement noise.

The manufacturer specifies a noise floor of $\leq -120\,\frac{\text{dBm}}{\text{Hz}}$ for the R&S ZNBT8 [86]. We terminate the stimulus port 1 with a $50\,\Omega$-impedance and measure the channel gain $|S_{21}(f)|^2$ at a tripod-mounted antenna connected to port 2, so that our measurement captures both ambient noise as well as noise of the instrument. In order to measure the instrument noise separately, we repeat the measurement with a $50\,\Omega$-impedance terminating both ports. Fig. 4.12 shows the frequency responses for different IF bandwidths. The results are averaged over 10 measurements each and with the CIRs time-windowed to 50 ns for clarity of illustration. Our observations are twofold: The noise floor in-

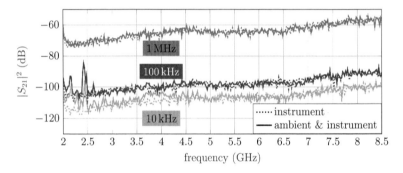

Figure 4.12.: VNA noise levels for different IF bandwidths

creases with larger IF bandwidth: For lower IF bandwidth range, the noise floor rises by $\approx 10\,\text{dB}$ for a tenfold bandwidth increase, which is expected. However, the noise is significantly stronger for an IF bandwidth of 1 MHz. Regarding the comparison of ambient and instrument noise, we observe that the high instrument noise floor for an IF bandwidth of 1 MHz is dominant. For $B_{\text{IF}} \leq 100\,\text{kHz}$, there is external interference exceeding the instrument noise floor below 2.5 GHz. This interference (Bluetooth and Wi-Fi transmissions from nearby devices) was present during all measurements. However, we cannot exclude the possibility that the absence of interference is due to a coincidental transmission break during the measurement of the respective frequency bins. As a worst-case estimate for the received interference from a Wi-Fi access point, we assume a Wi-Fi transmit power of 100 mW at 2.4 GHz at a distance of 5 m in free space, resulting in an interference power of $-34\,\text{dBm}$ at the antenna, which would be significantly above the measured instrument noise floor. In order to avoid strong interference in the 2.4 GHz-band and comply with the UWB standard, we use measurement data for frequencies above 3.1 GHz for the posture classification.

The measurement duration is approximately inversely proportionate to the IF bandwidth. With the multitude of channels, an IF bandwidth of 1 MHz is required to acquire the data for a posture snapshot within a reasonable time.

4.2.2 Relation to UWB regulations

In addition to a limitation of the average power spectral density of UWB signals to $\text{PSD}_{\text{UWB,max}} = -41.3 \frac{\text{dBm}}{\text{MHz}} = 74.1 \frac{\text{nW}}{\text{MHz}}$, the maximum permitted pulse power is limited to $P_{p,\text{max}} = 1\,\text{mW}$ measured over a bandwidth of 50 MHz [59]. Consequently, the maximum pulse energy is limited to $E_{p,\text{max,50 MHz}} = \frac{1\,\text{mW}}{50\,\text{MHz}} = 2 \cdot 10^{-11}\,\text{Ws}$ for every 50 MHz-subband of the occupied spectrum. For the following calculation, we will assume a flat spectrum of the UWB pulse. Thus, each subband of 1 MHz has the maximum permitted energy

$$E_{p,\text{max,1 MHz}} = \frac{E_{p,\text{max,50 MHz}}}{50} = 4 \cdot 10^{-13}\,\text{Ws}. \tag{4.8}$$

The dynamic range (i.e. the SNR) for a single pulse UWB measurement with maximum permitted pulse energy is hence equivalent to a VNA measurement with frequency bins of width 1 MHz and the same energy $E_{\text{bin}} = E_{p,\text{max,1 MHz}}$ per frequency bin.

$$\text{SNR}_{\text{bin}} = \frac{E_{\text{bin}}}{N_0 \cdot F_{\text{RX}}} = 80\,\text{dB} - F_{\text{RX,dB}}. \tag{4.9}$$

Consequently, the dynamic range of the VNA of $\approx 50 - 60\,\text{dB}$ (cf. Fig. 4.12 for $B_{\text{IF}} = 1\,\text{MHz}$ with a time window of 20 ns providing 17 dB gain) can be achieved with a single pulse UWB measurement with $F_{\text{RX,dB}} \leq 20\,\text{dB}$.

Time Windowing

UWB features very short pulses in the time domain. Thus, we can expect the received pulse to be confined to a short time interval (indoors typically $\leq 20\,\text{ns}$), i.e. a few time samples. In order to obtain the impulse response $\mathbf{h}$ of our measurements, the measured discrete frequency response $\mathbf{S}_{21}[\kappa \cdot \Delta f] = \mathbf{H}_n$ is transformed to the time domain using the inverse discrete Fourier transform (IDFT)

$$\mathbf{h}_k = \frac{1}{N_{\text{freq}}} \sum_{n=0}^{N_{\text{freq}}-1} \mathbf{H}_n \, e^{j2\pi \frac{k\,n}{N_{\text{freq}}}}. \tag{4.10}$$

The frequency domain measurements $\mathbf{H}_n$ are subject to the additive noise $\mathbf{H}_{n,\text{noise}}$ of variance σ_{freq}^2, which we assume to be i.i.d. The corresponding noise variance σ_t^2 of the noise affecting the transform result $\mathbf{h}_k$ in the time domain is hence

$$\mathbf{h}_{k,\text{noise}} = \frac{1}{N_{\text{freq}}} \sum_{n=0}^{N_{\text{freq}}-1} \mathbf{H}_{n,\text{noise}} \, e^{j2\pi \frac{k\,n}{N_{\text{freq}}}} \tag{4.11}$$

$$\sigma_t^2 = \frac{1}{N_{\text{freq}}^2} \, \sigma_{\text{freq}}^2 \, N_{\text{freq}} = \frac{\sigma_{\text{freq}}^2}{N_{\text{freq}}}, \tag{4.12}$$

where N_{freq} is the number of frequency bins. This is in line with Fig. 4.11, where $1.96 \cdot 10^{-9} = \sigma_t^2 = \frac{\sigma_{\text{freq}}^2}{N_{\text{freq}}} = \frac{1.27 \cdot 10^{-5}}{6501}$. Note that the entries of $\mathbf{h}_{k,\text{noise}}$ remain also i.i.d.

The frequency bin spacing determines the measurement duration in the time domain. With short CIRs due to limited multipath and short UWB pulses, most of the available N_{samp} time samples are pure noise. Hence, time-windowing the CIR can be used to improve the SNR. Let $K < N_{\text{samp}}$ be the number of time samples which contain a portion of the desired signal. For a time window length of K samples (i.e. limiting the CIR to its relevant part) and i.i.d time domain noise with variance σ_t^2, the noise variance in the frequency domain according to (4.12) is $\sigma_{\text{freq}}^2 = K \sigma_t^2$. With all the signal energy contained in the observation window, the SNR per bin improves by a factor of N_{samp}/K to

$$\text{SNR}_{\text{bin}} = \frac{E_{\text{bin}}}{N_0} \frac{N_{\text{samp}}}{K}. \tag{4.13}$$

With the time window of 20 ns applied for the antenna measurements in Section 4.1, the SNR per frequency bin is improved by 17 dB compared to the full measurement time of 1 μs.

SNR Reference and Bandwidth Dependency

In the remainder of this work, we use the transmit SNR as a reference for the analysis of transmit power dependencies. It is defined as

$$\mathrm{SNR_{TX}} = \frac{E_p}{\sigma_t^2}, \tag{4.14}$$

which is equivalent to the reception of a pulse with the entire pulse energy E_p confined to a single sampling instance. Note that this transmit SNR does not account for path loss (cf. Section 4.6.1). However, it provides a suitable base of comparison across different links as it scales directly with the transmit power. We scale the SNR, i.e. emulate different pulse energies, by scaling the noise variance σ_t^2.

When varying the signal bandwidth, additional frequency bins are considered in the evaluation, effectively using more energy for the measurement. As a result, the SNR increases proportionately to the bandwidth (i.e. to the number of frequency bins considered). This is in line with the UWB limit of the power spectral density, permitting higher transmit power for larger bandwidth. Consequently, a bandwidth increase by a factor of 10 comes with an increase of the reference SNR by 10 dB.

4.3 Measurement Procedure

4.3.1 Concurrent Multiport Measurements

Measurement equipment with more than two ports enables concurrent multiport measurements. Measuring multiple channels at the same time has several advantages: The most apparent benefit is the reduced measurement time. Furthermore, there is no need for manual or automated external switching between the different antennas, which significantly simplifies the data management process as data for one setup can be recorded and saved as whole. Consequently, the test subject can maintain more complex postures for the entire duration of the measurement. The resulting data for these postures is thus consistent, which is difficult to achieve with fewer ports than nodes, as rewiring takes too much time to keep a steady posture. An external switching board for a quick automated change of connections between device ports and antennas can mitigate this problem [11]. However, it is not trivial to design such a circuit for the given frequency range. Moreover, compensating for the signal alterations induced by the switching board itself increases the required calibration effort. Especially for large numbers of nodes like in our case, concurrent measurements are the only competitive approach for the data acquisition procedure.

Automating most of the configuration and measurement further improves the process. The VNA is configured and controlled remotely using Matlab. This allows for easy and largely automated measurements and optimization of the data management for further processing. For all measurements described in this chapter, data is first recorded as described and afterwards post-processed offline in Matlab. Data for every setup under test is bundled in a file containing the raw data from the device, the setup parameters, and the device configuration.

4.3.2   Measurement Process for a Single Posture Setup

We take 12 complete measurement snapshots comprising all links for each setup and test person. In order to cover possible variations of a setup, the posture is slightly modified in two different ways: During the measurement of three sets each, the test person moves within a range of a few centimeters (small scale variation). After three sets have been recorded, the person returns to a relaxed upright standing posture (no. 100, cf. Fig. 2.1) and takes a short pause before performing the measurement posture again in a modified way, e.g. with a different limb position within the range of the same posture for the acquisition of the next three sets (large scale variation). The "reset" to a relaxed posture in between measurement rounds and the conscious modification ensure a more representative variation in the 12 sets for each posture and setup.

4.4   Test Subjects

All posture measurements are conducted for three test subjects of different physique. Table 4.2 summarizes the respective information for test persons P1-P3. Despite the

test person	P1	P2	P3
sex	female	male	male
age	24 yrs.	29 yrs.	25 yrs.
height	1.58 m	1.78 m	1.90 m
weight	55 kg	70 kg	94 kg

Table 4.2.: Test subject data

similar age of the test subjects, the respective datasets are expected to differ in characteristics depending on the physique. In particular, the distances between nodes depend on the height and limb length, whereas the overall shape of the body influences the shadowing conditions and the wave propagation around the torso [88]. The variety in the human body characteristics covered with the three subjects enables both a comparison (for subject-dependent results) as well as a generalization (for subject-independent results) of the findings. In the remainder of this thesis, the test subject is noted whenever respective measurements are analyzed.

4.5 Environment

Fall prevention systems must operate reliably in a variety of environments from outdoor areas or inside public transport to indoor surroundings. Differences in the wireless channels of these environments, especially different multipath propagation, influence the measured on-body channels. An exhaustive analysis of all possible environments is not feasible. Consequently, we focus on indoor measurements in this work, as our target user group of old people spends the majority of time indoors. The environment for measurements is a furnished office, whose floor plan is shown in Fig. 4.13. The subject's and the VNA position vary for test person P1, P2, P3 and between the setups as displayed (open: black, clutter: blue).

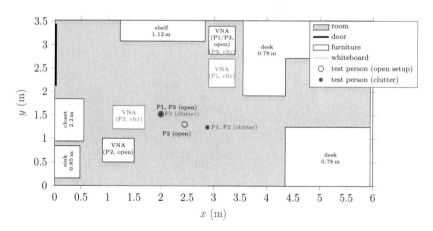

Figure 4.13.: Floor plan of the office environment

In order to account for changing indoor surroundings, we introduce variety in the measurements modifying the immediate surroundings within the same office room. In an *open setup*, the test subject performs all postures in the center of an empty space with a distance of at least 1.5 m to all reflecting items. In a *cluttered setup*, various reflecting and scattering items are placed in immediate vicinity of the test person, i.e. at a distance of not more than 1 m. Metallic sheets, tripods, a standing fan with a protecting cage, and the VNA placed on a rolling cart serve as scatterers and reflectors in various constellations and are subsequently referred to as "clutter". The clutter constellations differ between the subjects. The respective positions of the test subjects and the VNA are marked in Fig. 4.13. A selection of photographs of various setups is displayed in Fig. 4.14.

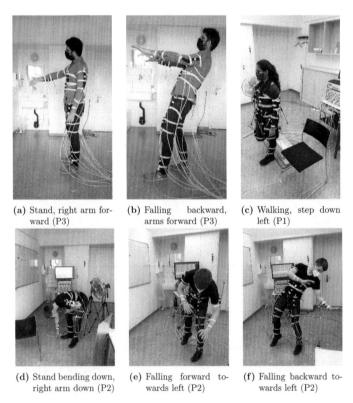

(a) Stand, right arm forward (P3)

(b) Falling backward, arms forward (P3)

(c) Walking, step down left (P1)

(d) Stand bending down, right arm down (P2)

(e) Falling forward towards left (P2)

(f) Falling backward towards left (P2)

Figure 4.14.: Photos of various measurement setups

setup		avg. path loss (dB) across all measurements
open	P1	-51.0 ± 7.5
	P2	-51.2 ± 7.4
	P3	-51.4 ± 7.3
cluttered	P1	-51.7 ± 7.2
	P2	-51.1 ± 7.3
	P3	-51.4 ± 7.3

Table 4.3.: Observed path loss for test subjects P1-P3

4.6 General Observations from Measured CIRs

Note: Parts of this section have been published in a very similar fashion in [36].

In the following, we will discuss various observations of the measured CIRs in order to give an intuition for the sets of measured CIRs. Thereby, CIRs of links which vary significantly across the postures are the backbone of the posture classification.

4.6.1 Path Loss

Table 4.3 shows the average path loss $\frac{E_{RX}}{E_{TX}}$ by setup and test subject. Note that this takes into account both LOS and NLOS links, explaining the strong variations (standard deviation $> 7\,\mathrm{dB}$). As it can be seen, the difference in the subject's physique does not significantly influence the average path loss over all links.

The observed path loss for a link depends on the distance between the respective nodes, the LOS conditions, and the orientation of the linearly polarized antennas. In the following, we take a closer look at different types of links. For this purpose, we group the nodes by their position, namely on the arms, on the legs, on the front of the torso, or on the back. Consequently, we can analyze links between two of these groups, e.g. from the leg-mounted nodes to the front of the torso. Intuitively, links with a strong variability across the postures are more helpful for the posture distinction. Fig. 4.15 shows the empirical cumulative distribution functions (CDFs) for the measured average path loss for links between different node positions for all postures of test subject P2. We observe that links from the limbs to the torso exhibit greater path loss variation. Links between the static nodes around the torso are significantly more stable, which is expected. It is noteworthy that Fig. 4.15 includes the effects of path loss variations within the same posture, as well as path loss variations between

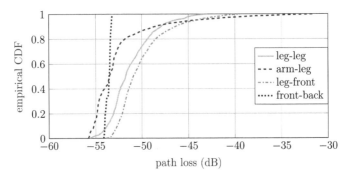

Figure 4.15.: Path loss CDF per link type (P2, open setup)

different postures. While the former is ideally negligible, the latter should be large to facilitate discriminability of the postures. An assessment of the path loss and the resulting received signal energy as a distinctive feature is provided in Chapter 6.

4.6.2 Line of Sight / Non-Line of Sight

Next, we will examine typical LOS and NLOS characteristics. It has been shown that the LOS conditions of on-body links have a significant influence on the channel [27,88]. Consequently, on-body links with a LOS in some postures and NLOS in others will contribute to the discriminability of the respective postures. In contrast, links between nodes on the torso face more static, i.e. posture-independent channels amongst each other. Their contribution to the discriminability is thus expected to be lower.

For the purpose of illustration, we take a closer look at the link between the right wrist (node 6) and the right ankle (node 18) for different postures, which are shown in Fig. 4.16.

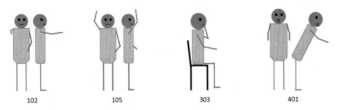

Figure 4.16.: Selected postures for LOS/NLOS analysis

Fig. 4.17 shows the magnitude of the respective CIR captured without items in the immediate surroundings (open setup) for test subject P3.

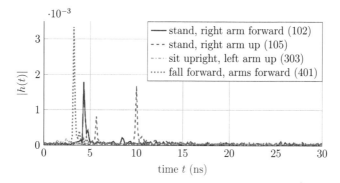

Figure 4.17.: Magnitude of measured CIRs for wrist-ankle link

We observe that the characteristics of this link vary significantly between the postures: A strong LOS path without additional MPCs characterizes the channel for postures 102 (stand, arm forward) and 401 (fall forward, arms forward). The LOS path also occurs for posture 105 (stand, arm up), but here we observe another MPC in addition. From the delay and the conditions during measurement, this MPC is the reflection from the metallic whiteboard mounted on the office wall, which accounts for a very strong reflection. The larger amplitude despite the longer path can be explained by the antenna patterns. The LOS component lies in the direction of a minimum of the radiation patterns of both antennas, whereas the reflector faces a direction close to the maximum. In case of posture 303 (sitting, arm up), the LOS is very weak, and the link exhibits the characteristics of a NLOS connection.

For an intuitive and straightforward comparison, a closer look at the cumulative energy is a suitable approach. Fig. 4.18 shows the cumulative energy for the time window of the first 30 ns, both the absolute value (Fig. 4.18a) as well as normalized to the respective maxima at $t = 30$ ns (Fig. 4.18b).

The pulse arrival times with the respective strong increase of the cumulative energy are clearly visible in both figures. We observe from Fig. 4.18b that the LOS peaks of postures 102 and 401 contain more than 80 % of the total energy in the displayed time window. For posture 105, the two MPCs together are responsible for more than 80 % of the total energy. This illustration further reveals a weak MPC for posture 303

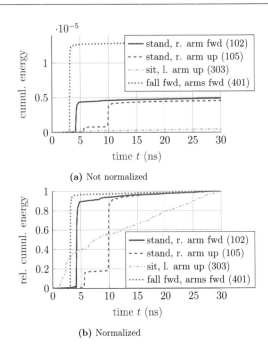

(a) Not normalized

(b) Normalized

Figure 4.18.: Cumulative energy over a time window

at $t \approx 3\,\text{ns}$ with about $25\,\%$ of the total energy, which is not as clearly visible in the previous plots. From Fig. 4.18a we further observe that the receiving nodes are subject to the same noise floor, as the slope of the curves is identical for $t > 15\,\text{ns}$ (in close range of the noise variance in Fig. 4.11a), where the energy increase is only due to the integration of the noise. Note that the slope differs in Fig. 4.18b only due to the normalization.

4.6.3 Surrounding Objects

In order to get an intuition for the influence of the environment, we take a closer look at the standing posture 101 performed by test subject P1 in different setups as depicted in Fig. 4.19. Measurements for this posture cover a variety of setups with various potentially reflecting objects at different distances in front of the leg, which makes it the posture of choice for this analysis. In particular, we are interested in

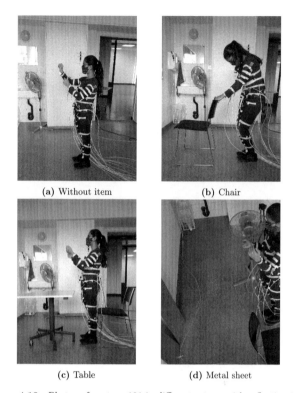

(a) Without item   (b) Chair

(c) Table   (d) Metal sheet

Figure 4.19.: Photos of posture 101 in different setups with reflecting items

the degree to which these items introduce additional MPCs. The link between left knee and left ankle is examined for illustrative purposes, as we can expect to observe additional MPCs from the reflecting items for this link. Fig. 4.20 shows the cumulative link energy as introduced before.

It can be observed that the metallic chair frame does not cause a visible second MPC, and the CIR is very similar to the case without any items around. A table with a thicker metallic leg provides a weak second path with a delay of approximately 3 ns in addition to the LOS. A vertical metal sheet leaning to a wall acts as a strong reflector, which induces a stronger second MPC with a path delay of approximately 7 ns. This result is well in line with the expectations of weaker reflections from the round legs of furniture and a strong echo from the metal plane. The differences in the LOS peak

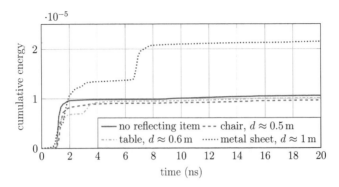

Figure 4.20.: Cumulative energy for knee–ankle link (standing posture 101)

amplitudes are likely due to variations of the antenna placement, as the measurements of the metal sheet setup were taken during a different session than the other three setups. Overall we can conclude that the effects of the environment are observable in the measurements, which underlines the need to take different setups into account. The effect of the human body itself, however, is still dominant over the environmental influence [89].

4.7   Summary

This chapter provided details on the measurement process, including the environment, the equipment and the various steps of the data acquisition procedure. We have further outlined some characteristics of the measured CIRs: Depending on the selected link and the environment, we observe the presence of different MPCs for the same link. Depending on the link distance and LOS conditions, the observed path loss varies between postures, which exhibits potential to be used for posture classification.

5

Maximum Likelihood Performance Analysis with Simplified Signal Model

In this chapter we introduce a simplified system model, based on which we provide the maximum likelihood classifier for the three levels of raw data categorized in Chapter 4. The simplified model shall serve for an evaluation of the expected feasibility of the posture classification task based on the acquired data. This chapter follows the procedure described in [10, 90].

In the remainder of this chapter, the following notation is used: The set of all 43 postures we aim to recognize is denoted as $\mathcal{P}$. For each posture $p \in \mathcal{P}$ we have a set $\mathcal{S}^{(p)} = \{\mathbf{s}^{(p,1)}, ..., \mathbf{s}^{(p,Q_p)}\}$ of measurement snapshots $\mathbf{s}^{(p,q)}$, which are a concatenation of the complex CIR vectors $\mathbf{h}^{(p,q,l)} \in \mathbb{C}^{N_{\text{samp}} \times 1}$ for all links $l \in N_{\text{link}}$, which each contain N_{samp} time samples:

$$\mathbf{s}^{(p,q)} = \begin{bmatrix} \mathbf{h}^{(p,q,1)} \\ \mathbf{h}^{(p,q,2)} \\ \vdots \\ \mathbf{h}^{(p,q,N_{\text{link}})} \end{bmatrix} \in \mathbb{C}^{N_{\text{link}} \cdot N_{\text{samp}} \times 1}. \tag{5.1}$$

5.1 Simplified Signal Model

Despite the high SNR of the VNA measurements, the acquired data contains various noise components from different sources. Posture variations add randomness at different places, such as amplitude variations of the CIRs due to polarization mismatch, or delay and amplitude differences due to varying agent-anchor distances. As the distributions of these disturbances are unknown and therefore difficult to model accurately, we will rely on the following simplified system and signal model.

We assume all VNA measurements $\mathbf{s}^{(p,q)}$ as noise-free, i.e. we do not consider posture variations outside the scope of our set of measurements. Despite this simplification, a variety of realizations of each posture is covered as described in Chapter 4. Furthermore, the measurement noise of the VNA is considered as a part of the signal, too. The focus of our evaluation is instead placed on a variation of the transmit power of the on-body nodes, which is a crucial parameter for a battery-powered system. In order to emulate a variation of the transmit power, we scale the SNR of the measured CIRs accordingly. This is done by adding i.i.d. complex additive white Gaussian noise (AWGN) with an appropriately chosen variance to all CIRs $\mathbf{h}^{(p,q,l)}$, so that we have the noisy measurements

$$\mathbf{m}^{(p,q)} = \mathbf{s}^{(p,q)} + \mathbf{n}, \tag{5.2}$$

where $\mathbf{n}$ is circularly symmetric complex Gaussian distributed:

$$\mathbf{n} \sim \mathcal{CN}\left(\mathbf{0}, \sigma^2\,\mathbf{I}_{N_{\text{link}} \cdot N_{\text{samp}}}\right) \tag{5.3}$$

Consequently, the noise-free posture realizations $\mathbf{s}^{(p,q)}$ are the individual mean values for a Gaussian mixture model, so that for a given posture p the randomly generated measurement samples are distributed according to the conditional PDF

$$f(\mathbf{m}|p) = \frac{1}{Q_p} \sum_{q=1}^{Q_p} f_N\left(\mathbf{m}; \mathbf{s}^{(p,q)}, \sigma^2 \mathbf{I}\right). \tag{5.4}$$

The maximum likelihood estimate of the posture corresponding to measurement $\mathbf{m}$

$$\hat{p} = \arg\max_p \mathcal{L}_p = \arg\max_p f(\mathbf{m}|p) \tag{5.5}$$

is also the maximum a posteriori solution for equiprobable posture realizations $\mathbf{s}^{(p,q)}$ for all postures p.

Choosing the Noise Variance

In order to justify the assumption of noisefree VNA measurements, the artificial noise for emulation of lower transmit powers is chosen significantly stronger than the actual measurement noise in the VNA data. We select the artificial noise variance $\sigma^2_{\text{artif.}}$ based on the largest variance σ^2_{meas} of measurement samples in all CIRs $\mathbf{h}^{(p,q,l)}$ for all postures p, all links l, and all measured snapshots q. For each CIR $\mathbf{h}^{(p,q,l)}$, σ^2_{meas} is evaluated for a time window $t = [100\,\text{ns}, 900\,\text{ns}]$ without any measurable multipath components (delay spread $\approx 30\,\text{ns}$). For all evaluations, the additional artificial noise variance is chosen at least a magnitude larger than the maximum measured variance across all CIRs

$$\sigma^2_{\text{artif.}} \geq 10 \max_{p,q,l}(\sigma^2_{\text{meas}}), \tag{5.6}$$

so that $\sigma^2 = \sigma^2_{\text{meas}} + \sigma^2_{\text{artif.}} \approx \sigma^2_{\text{artif.}}$. All evaluations with additional artificial noise are performed over multiple runs with different artificial noise realizations, with the reported results being the average over all runs.

5.2 Maximum Likelihood Classifier

Note: Detailed derivations with intermediate steps are provided in Appendix A.

5.2.1 Complex CIR

A coherent receiver directly measures the superposition of complex CIR and the complex noise components. We define the decision variable $\mathbf{d}_1$ as a scaled version of the measurement vector:

$$\mathbf{d}_1 = \frac{\sqrt{2}}{\sigma}\mathbf{m} \qquad \sim \mathcal{CN}\left(\frac{\sqrt{2}}{\sigma}\mathbf{s}^{(p,q)}, 2\,\mathbf{I}_{N_{\text{link}} \cdot N_{\text{samp}}}\right). \tag{5.7}$$

Consequently, its real part $\mathbf{d}_{1,I}$ and its imaginary part $\mathbf{d}_{1,Q}$ are standard normally distributed. For notational convenience we define

$$\widetilde{\mathbf{s}^{(p,q)}} = \frac{\sqrt{2}}{\sigma}\mathbf{s}^{(p,q)}, \tag{5.8}$$

so that we have the conditional PDF

$$f(\mathbf{d}_1|p) = \frac{1}{Q_p} \sum_{q=1}^{Q_p} f_N\left(\mathbf{d}_1; \widetilde{\mathbf{s}^{(p,q)}}, 2\mathbf{I}\right). \tag{5.9}$$

The maximum likelihood decision is therefore

$$\hat{p} = \arg\max_{p} \mathcal{L}_p = \arg\max_{p} \frac{1}{Q_p} \sum_{q=1}^{Q_p} \exp\left(-\frac{1}{2}\left\|\mathbf{d}_1 - \widetilde{\mathbf{s}^{(p,q)}}\right\|^2\right). \tag{5.10}$$

Note that the sum over Q_p different snapshots for each posture prevents the use of the log-likelihood for simplification.

5.2.2 Magnitude of CIR

As outlined in Chapter 3, information about the signal phase may not be available depending on the hardware implementation of the receiving nodes. The measured magnitude of the CIRs is Rician distributed with the posture snapshots $\mathbf{s}^{(p,q)}$ as respective means. We define our decision variable $\mathbf{d}_2$ based on the squared magnitude of the CIR, so that we have

$$\mathbf{d}_2 = \mathbf{d}_1 \odot \mathbf{d}_1^* = \frac{2}{\sigma^2}\,\mathbf{m} \odot \mathbf{m}^* \qquad \sim \chi^2\left(2, \lambda_2 = \widetilde{\mathbf{s}^{(p,q)}} \odot \widetilde{\mathbf{s}^{(p,q)}}^*\right), \tag{5.11}$$

where the $\odot$-operator denotes the element-wise Hadamard multiplication of the vectors. For the distribution of $\mathbf{d}_2$

$$f(\mathbf{d}_2|p) = \frac{1}{Q_p} \sum_{q=1}^{Q_p} f_{\chi^2}\left(\mathbf{d}_2; 2, \lambda_2^{(p,q)}\right) \tag{5.12}$$

we obtain the maximum likelihood decision

$$\hat{p} = \arg\max_{p} \frac{1}{Q_p} \sum_{q=1}^{Q_p} \exp\left(-\frac{1}{2}\widetilde{\mathbf{s}^{(p,q)}}^H \widetilde{\mathbf{s}^{(p,q)}}\right) \prod_{k=1}^{N_{\text{link}} \cdot N_{\text{samp}}} I_0\left(\sqrt{|s_k^{(p,q)}|^2\, d_{2,k}}\right), \tag{5.13}$$

where $I_0(\cdot)$ denotes the modified Bessel function of the first kind with order zero.

5.2.3 Signal Strength

In this section, the received signal strength for each link is measured separately, the respective measurement for the CIR $\mathbf{h}^{(p,q,l)}$ of a link is denoted as $\mathbf{m}^{(\cdot,\cdot,l)} \in \mathbb{C}^{N_{\text{samp}} \times 1}$, so that in agreement with (5.1) we have

$$\mathbf{m}^{(\cdot,\cdot,l)} = \mathbf{h}^{(p,q,l)} + \mathbf{n}^{(\cdot,\cdot,l)} \tag{5.14}$$

In this case, the decision variable vector $\mathbf{d}_3$

$$\mathbf{d}_3 = \begin{bmatrix} \mathbf{1}_{N_{\text{samp}}}^T & \mathbf{0} & \cdots\cdots\cdots & \mathbf{0} \\ \mathbf{0} & \mathbf{1}_{N_{\text{samp}}}^T & \mathbf{0} & \cdots\cdots & \vdots \\ \vdots & \ddots & \ddots & \ddots & \vdots \\ \mathbf{0} & \cdots\cdots\cdots\cdots & \mathbf{0} & \mathbf{1}_{N_{\text{samp}}}^T \end{bmatrix} \mathbf{d}_2 = \frac{2}{\sigma^2} \begin{bmatrix} \left(\mathbf{m}^{(\cdot,\cdot,1)}\right)^H \mathbf{m}^{(\cdot,\cdot,1)} \\ \left(\mathbf{m}^{(\cdot,\cdot,2)}\right)^H \mathbf{m}^{(\cdot,\cdot,2)} \\ \vdots \\ \left(\mathbf{m}^{(\cdot,\cdot,N_{\text{link}})}\right)^H \mathbf{m}^{(\cdot,\cdot,N_{\text{link}})} \end{bmatrix} \tag{5.15}$$

has N_{link} entries, which are determined by the received signal strength (RSS) values of the respective links. With

$$\mathbf{d}_3 \sim \chi^2 \left(2N_{\text{samp}}, \lambda_3^{(p,q)} = \frac{2}{\sigma^2} \begin{bmatrix} \left(\mathbf{h}^{(p,q,1)}\right)^H \mathbf{h}^{(p,q,1)} \\ \vdots \\ \left(\mathbf{h}^{(p,q,N_{\text{link}})}\right)^H \mathbf{h}^{(p,q,N_{\text{link}})} \end{bmatrix} \right) \tag{5.16}$$

we have

$$f(\mathbf{d}_3|p) = \frac{1}{Q_p} \sum_{q=1}^{Q_p} f_{\chi^2}\left(\mathbf{d}_3; 2\,N_{\text{samp}}, \lambda_3^{(p,q)}\right) \tag{5.17}$$

and the maximum likelihood decision is made according to

$$\hat{p} = \arg\max_p \frac{1}{Q_p} \sum_{q=1}^{Q_p} \exp\left(-\frac{1}{2}\,\widetilde{\mathbf{s}^{(p,q)}}^H \widetilde{\mathbf{s}^{(p,q)}}\right) \prod_{l=1}^{N_{\text{link}}} \left(\frac{d_{3,l}}{\lambda_{3,l}^{(p,q)}}\right)^{(N_{\text{samp}}-1)/2} I_{N_{\text{samp}}-1}\left(\sqrt{\lambda_{3,l}^{(p,q)}\,d_{3,l}}\right). \tag{5.18}$$

Numerical evaluation of the high-order Bessel functions can be problematic [91]. However, a noncentral χ^2-distribution with noncentrality parameter λ and high degrees of freedom $\nu = 2\,N_{\text{samp}}$ can be well approximated with a Gaussian distribution [92] with mean $\mu_{\text{approx.}} = \nu + \lambda$ and variance $\sigma^2_{\text{approx.}} = 2\nu + 4\lambda$. As the degrees of freedom are determined by the number of time sampling instances N_{samp}, a sufficiently high sampling rate permits this approximation. In our case $N_{\text{samp}} \gg 100$, so that the approximate maximum likelihood decision is made according to

$$\hat{p}_{\text{approx}} = \arg \max_p \frac{1}{Q_p} \sum_{q=1}^{Q_p} \prod_{l=1}^{N_{\text{link}}} \exp \left(-\frac{\left(d_{3,l} - \mu^{(p,q)}_{\text{approx},l} \right)^2}{2 \left(\sigma^{(p,q)}_{\text{approx},l} \right)^2} \right) \tag{5.19}$$

$$= \arg \max_p \frac{1}{Q_p} \sum_{q=1}^{Q_p} \exp \left(-\frac{1}{2} \sum_{l=1}^{N_{\text{link}}} \frac{\left(d_{3,l} - \mu^{(p,q)}_{\text{approx},l} \right)^2}{\left(\sigma^{(p,q)}_{\text{approx},l} \right)^2} \right) \tag{5.20}$$

with $\mu^{(p,q)}_{\text{approx},l} = 2\,N_{\text{samp}} + \lambda^{(p,q)}_l$ and $\sigma^{(p,q)}_{\text{approx},l} = 4\,N_{\text{samp}} + 4\,\lambda^{(p,q)}_l$. This approximation is applied in cases where a numerical evaluation of the Bessel function fails.

5.3 Feasibility Assessment

In the following, we aim to determine the SNR range in which the classification task is feasible for the available and artificially noisy data. For all evaluations, we limit the CIR to a time window of $T_{\text{win}} = 30\,\text{ns}$, which is the longest observed delay spread. Consequently, the measured CIRs include all relevant MPCs. Unless noted otherwise, the UWB range $(3.1\,\text{GHz} - 8.5\,\text{GHz})$ is considered, i.e. a bandwidth of $B = 5.4\,\text{GHz}$ around $f_c = 5.8\,\text{GHz}$. We further assume perfect time synchronization.

5.3.1 Discussion of SNR Dependency

With full knowledge about the noise distribution (i.e. without splitting the posture realizations into calibration and test data), the maximum likelihood classifier is the achievable optimal classifier. Fig. 5.1 shows the accuracy for a variation of the transmit SNR as defined in (4.14).

For the coherent case, we observe a reliable classification with an accuracy above 0.9 for an SNR above 32 dB. Without phase information, the same accuracy requires an SNR increase by a decade. The signal energy-based approach is the least robust with a significantly weaker performance, requiring an SNR of 52 dB for an accuracy of 0.9. Thus, a raw data level containing more information allows to reduce the transmit

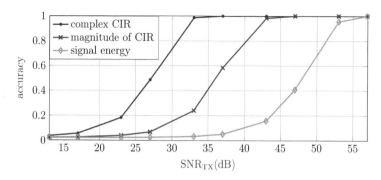

Figure 5.1.: Maximum likelihood accuracy

power by a factor of 10 compared to the next lower raw data level containing less information.

The pairwise inter-sample distances between snapshots of different postures determine the pairwise error probabilities: A misclassification occurs when the noise is large enough to induce a crossing of the decision boundary. Hence, a more detailed examination of the inter-sample distances provides a better understanding of the SNR dependency. We compute the squared inter-sample distances $d^2_{p_m,p_n,q_i,q_j,\cdot}$ between snapshots q_i, q_j of different postures $p_m \neq p_k$ as

$$d^2_{p_m,p_n,q_i,q_j,1} = \left\| \mathbf{s}^{(p_m,q_i)} - \mathbf{s}^{(p_n,q_j)} \right\|^2 \tag{5.21}$$

$$= \left(\mathbf{s}^{(p_m,q_i)} - \mathbf{s}^{(p_n,q_j)} \right)^H \left(\mathbf{s}^{(p_m,q_i)} - \mathbf{s}^{(p_n,q_j)} \right) \tag{5.22}$$

$$d^2_{p_m,p_n,q_i,q_j,2} = \sum_{k=1}^{N_{\text{samp}}} \left(|\mathbf{s}_k^{(p_m,q_i)}| - |\mathbf{s}_k^{(p_n,q_j)}| \right)^2 \tag{5.23}$$

$$d^2_{p_m,p_n,q_i,q_j,3} = \sum_{l=1}^{N_{\text{link}}} \left| \left\| \mathbf{h}^{(p_m,q_i,l)} \right\|^2 - \left\| \mathbf{h}^{(p_k,q_j,l)} \right\|^2 \right| \tag{5.24}$$

for the three raw data levels, respectively. Using the data for test subject P2 for illustration, we compute the squared Euclidean distance between all pairs of measurements, i.e. for all postures and environments. Classification errors occur most likely for the postures with the shortest distance to each other. Every CIR is windowed to a length of 30 ns, thus removing the purely noisy part without MPCs.

Fig. 5.2 shows these squared distances for all three raw data levels as boxplots, where the top and bottom of the boxes mark the third and first quartile, respectively. The median squared distance to samples of other postures is marked vertically inside the boxes, and the whiskers reach the minimum and maximum squared distances.

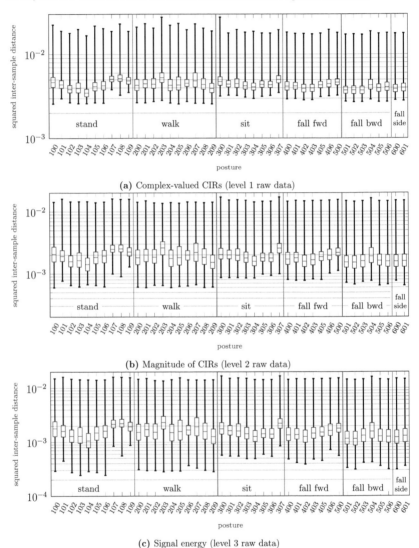

(a) Complex-valued CIRs (level 1 raw data)

(b) Magnitude of CIRs (level 2 raw data)

(c) Signal energy (level 3 raw data)

Figure 5.2.: Inter-sample distances between different postures

Challenging our simplifying assumption of noiseless VNA measurements, we further check the hypothetical inter-sample distance based solely on the measurement noise for comparison. Shifting the observation window of 30 ns to a time beyond the delay spread, measurement noise vectors are obtained which do not contain any stimulus response. Of the squared inter-sample distances computed from these noise vectors, 95 % are shorter than $5 \cdot 10^{-5}$ (complex CIR), 10^{-5} (magnitude), and $3 \cdot 10^{-6}$ (signal energy), respectively. With a difference of the signal- and noise-based inter-sample distances of two orders of magnitude, we conclude that the SNR of the measurement is sufficiently high to justify the assumption of noise-free VNA measurements for the simplified signal model.

From Fig. 5.2 we see that with comparatively larger minimal inter-sample distances, postures 107-109 (stand, bending down) and 307 (sit, bending down) are least likely to be misclassified due to Gaussian noise. This is not surprising, as these postures are visibly distinct and their characteristics thus deviate significantly from other postures. In general, the sitting postures are slightly more distinct than the postures of the other groups, as can be seen from the larger minimal distances to other postures. In contrast, the walking postures (2xx) have small minimal distances between their samples, as their differences are mostly limited to the relative positions of the nodes on the legs. As these differences are only observed in few of the 153 links, their influence in the very high-dimensional space (where distances are computed) is significantly reduced.

In order to illustrate this dependency between link selection and intra-posture distance, we take a look at the subspace spanned by only the links between the four nodes on the legs (i.e. on the knees and ankles). Fig. 5.3 shows the minimal squared distances between the postures in this subspace, i.e. $d^2_{\min,p_0,\cdot} = \min_{p \neq p_0, q_i, q_j} d^2_{p_0,p,q_i,q_j,\cdot}$ while only considering links between the leg-mounted nodes. In contrast to the observations in Fig. 5.2, walking postures have a larger inter-sample distance, and sitting postures are more likely to be misclassified due to their similar orientation of the test subject's legs across most sitting postures, which is reflected in their small minimal distances. Hence, different links are characteristic for different groups of postures. In Chapter 7 we further discuss suitable selection methods for the links and on-body nodes.

5.3.2 Maximum Likelihood-Inspired Bayes Classifier

Following the assumptions of the simplified signal model in Section 5.1, we have so far limited the possible posture realizations to the ones which are known to the clas-

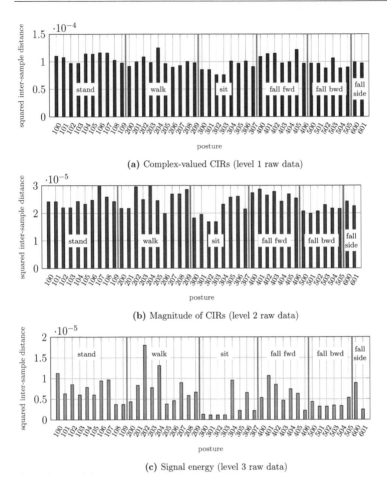

(a) Complex-valued CIRs (level 1 raw data)

(b) Magnitude of CIRs (level 2 raw data)

(c) Signal energy (level 3 raw data)

Figure 5.3.: Shortest inter-sample distances between different postures based on leg-mounted nodes

sifier. This assumption does not hold in practical applications where it is impossible to calibrate the system with all possible posture realizations. Therefore, we will now drop this simplification and split the 12 measurements per setup (posture, test subject, environment) into calibration and test snapshots at random. While this approach provides a more realistic evaluation of the classification accuracy in real-world scenarios, it is no longer the theoretically optimal classifier, as the differences between snapshots of the same setup contain variations which are not captured in the AWGN model.

Fig. 5.4 shows the accuracy for a calibration/test split of 2:1. None of the approaches

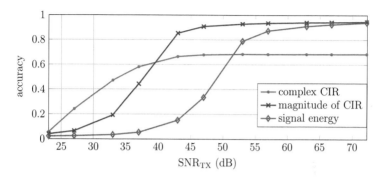

Figure 5.4.: Accuracy of a maximum likelihood-inspired Bayes classifier

achieves 100 % accuracy, and the performance is inferior to the maximum likelihood approach with the simplified signal model (cf. Fig. 5.1). This is expected due to the limitations from the simplification by the model. In particular, not all differences between realizations of the same posture are covered by the Gaussian assumption, which results in test samples being misclassified as noisy realizations of a different posture. Moreover, the behavior for classification based on the complex-valued CIR is striking, as the maximum accuracy for lower noise variances does not exceed 0.7, whereas both other approaches achieve an accuracy of 0.95 (magnitude of the CIR) and 0.9 (signal energy), respectively. The latter approaches are comparatively robust towards minor variations of the posture as the amplitude of the CIRs does not change significantly with slight movements of the test person. However, these minor variations potentially induce a significant change in the phase: With a center frequency of 5.8 GHz, a distance variation of $\frac{\lambda}{2} \approx 2.6$ cm induces a 180°-phase shift for the respective link. Posture realizations covering movements within that range exhibit a large Euclidean distance in the complex signal space due to their phase difference. As a consequence, the decision regions from the training data are no longer accurate for the test data, leading to a lower classification accuracy. The amplitude difference, however, is significantly smaller for such minor variations, leaving the magnitude- and signal energy-based classifiers less affected by the split into train and test data. The strong performance of the magnitude-based approach supports the conjecture of the phase differences as the limiting factor in the coherent case.

5.3.3   Impact of Frequency Range

The frequency range is generally a crucial parameter of wireless systems, as it influences the hardware complexity, antenna dimensions, time resolution, permitted transmit power, and potential external interference. Thus, it is standing to reason to evaluate the influence of the frequency range of the maximum likelihood classification performance. For this analysis, the bandwidth is varied from 5 MHz to 5.4 GHz. The lower edge of the frequency band is fixed to $f_{min} = 3.1$ GHz. Fig. 5.5 illustrates the bandwidth dependency of the maximum likelihood classifiers. The transmit SNR according to (4.14) for the full bandwidth is set to $\mathrm{SNR_{TX}} = 52$ dB. With the use of additional

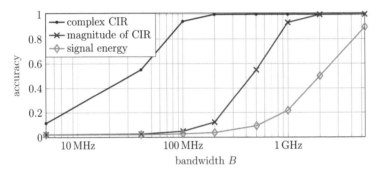

Figure 5.5.: Maximum likelihood accuracy vs. bandwidth

frequency bins for the measurement, the measurement energy increases proportionately to the bandwidth. This is in line with the UWB regulations limiting the transmit power spectral density, permitting higher transmit power for larger bandwidth.

We observe an increased accuracy with increasing bandwidth for all three raw data levels. With complex CIRs as raw data, a classification accuracy above 0.9 can be achieved with just 100 MHz bandwidth. For the same accuracy, a bandwidth of $B \geq 1$ GHz is required for noncoherently received CIRs. The signal energy-based classification requires the full bandwidth of 5.4 GHz for the same accuracy.

In order to revisit the underlying phenomena beyond the increase in transmit power, we look at the effect of a bandwidth increase for the CIR measurement at each raw data level. For a narrowband signal, the measured CIR is a single complex-valued channel coefficient. With a small change in the respective link due to a change of posture, its amplitude varies little compared to its phase: A path length difference of less than 3 cm introduces a phase shift of more than 180°, but only a minor amplitude fluctuation. Hence, such changes of the body posture are detected easier in complex-valued raw data for a small bandwidth. With increasing bandwidth, the MPCs and their individual delays become more prominent and distinct, which explains the increase in accuracy for the magnitude-based classifier: With 1 GHz bandwidth, a time resolution of approx. 1 ns and thus a spatial resolution of 30 cm is achieved, allowing to separate different MPCs such as the LOS and the ground reflection path. This, however, does not apply to the signal energy-based classifier, which benefits from frequency diversity only through more stable measured energy values of each link. A detailed individual analysis for our classifier of choice is conducted in Section 7.1.

5.4 Summary

In this chapter, we have introduced the maximum likelihood classifiers for the three raw data levels introduced in Chapter 3 within the frame of a simplified system model, which assumes the VNA measurements from Chapter 4 as noise-free. From the classification accuracy of the maximum likelihood classifiers we observed a complexity-robustness tradeoff: Each increase of the raw data complexity relaxes the SNR requirement for a fixed accuracy threshold by about a decade. With a split into training and test data, the simplified model no longer applies. As a result, classification based on the complex-valued CIR is inferior due to phase variations, whereas the magnitude and signal energy prove to be more robust in the high SNR regime. The classification accuracy generally benefits from larger signal bandwidth, especially for lower raw data complexity.

6

Classifier Selection from Common Machine Learning Approaches

After the previous feasibility evaluation based on the maximum likelihood classifiers for the simplified signal model, this chapter approaches the classification problem from an application-oriented direction. Machine learning has become the go-to standard for most classification tasks nowadays, with more advanced methods being developed constantly. For the case at hand, however, we rely on established approaches which we see merely as tools, without the ambition to conduct machine learning research. A brief introduction to the terminology and basic concepts of the machine learning space can be found in [46]. More detailed explanations are provided in [45] and [93]. The following introductions in this chapter are based on these works.

6.1 Formulation of the Classification Task as a Machine Learning Problem

The core of a classification task is the assignment of a *class label* (in our case: a posture or a group) to a *data point* (also referred to as *sample*, in our case: snapshot measurement) based on certain characteristic *features* of the data point (e.g. signal energy of on-body links). While these characteristics are not necessarily physically meaningful or human-interpretable, the problem at hand has a variety of them, such as well-known channel characteristics like time of flight, delay spread, maximum amplitude, number of MPCs, etc. The classification can be seen as a function which takes the features of a data point as input and provides a class label estimate as output. Our data is labeled, i.e. every data point has a true class it belongs to (*ground truth*). As the class labels are known for the training data and used for learning the decision boundaries, our approaches fall in the domain of *supervised learning*. Based on the labeled training data, the classifier (also referred to as *model*) learns decision boundaries between the classes and assigns the most likely class label to an unseen test data point. The learning process of the class boundaries is an optimization problem to minimize a loss (such as the number of misclassifications in the training data). Parameters of this optimization depend on the classifier and are referred to as *hyperparameters*, which have to be specified by the designer and are not learnt from the data. The procedure to find suitable hyperparameters is explained in Section 6.2.3.

The goal of this chapter is the systematic selection of feature-classifier-combinations which are suitable for the given posture classification problem. We pick from established approaches based on the different nature of the available raw data as introduced in Section 3.3, and select appropriate hyperparameters and features in a systematic way. All evaluations are implemented in Python using the `scikit-learn` library [94, 95], which provides implementations of all classifiers used, hyperparameter tuning functions, etc.

6.2   Data Space & Preprocessing

In order to gain an intuition of the data space this work operates in, we use t-distributed Stochastic Neighbor Embedding (tSNE) [96], a nonlinear dimensionality reduction approach, to illustrate the posture clusters in a two-dimensional space. While the observed clusters can give an intuition about the underlying high-dimensional data, the output of the dimensionality reduction depends on the parametrization and does not translate directly to the actual data sample distances. Fig. 6.1 shows a two-dimensional tSNE representation of the data based on the magnitude of the CIRs (Fig. 6.1a) and the signal energy measurements (Fig. 6.1b). For clarity of illustration, only data for test subject 2 is displayed. The marker type indicates the posture, the colors indicate the respective group.

For the case of the magnitude of CIRs in Fig. 6.1a, we observe that postures of the "sit" group (green) appear separate from the other group clusters, which are overlapping. This is intuitive as the nature of sitting postures is different from standing, walking or falling, which are mostly upright postures on one or both legs. We further observe that multiple measurements of the same posture (same marker and same color) are clustered comparatively densely, which is expected as the differences are only due to conscious and unconscious variations of the same posture. The extent of this variation differs, as we can for instance see from the measurements of "stand, bend down w/ left

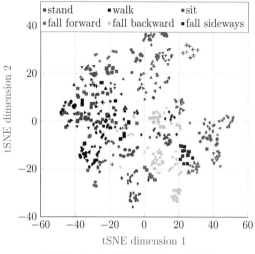

(a) Magnitude of CIRs (level 2 raw data)

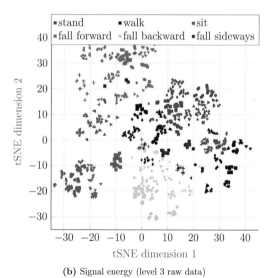

(b) Signal energy (level 3 raw data)

Figure 6.1.: tSNE visualization of data for test subject 2

arm stretched" (blue "+", top right) and "walk, step down left" (red "+", scattered in center). Note that when using the time sequences of the CIR time samples as input vectors for the tSNE method, a time shift of a spiky CIR by a single sample can induce a very large distance between the vectors. Thus, the tSNE illustrations need to be considered with caution.

From the signal energy-based tSNE illustration in Fig. 6.1b, we observe that the postures are also visibly clustered. Furthermore, the groups (indicated by color) tend to be clustered as well, except for the walking postures (red). Unlike the CIRs in the previous case, energy values are independent of small delay differences. Furthermore, the lower dimensionality of the raw data (N_{link} instead of $N_{\text{samp}} \cdot N_{\text{link}}$) helps to make the tSNE illustration more accurate as it suffers less from the curse of dimensionality (i.e. similar distances between points in very high-dimensional spaces).

6.2.1   Splitting

The set of available data points is split into a training set and a test set at random with the relation of $2 : 1$ in favor of the training set. The split is performed in such a way that the proportion of samples is identical with regard to the postures, environments, and test subjects in both sets. For the robustness analysis in Chapter 7, the splits are adapted to evaluate the influence of the data selection.

6.2.2   Cross-Validation

With a large amount of data available, it is common practice to split the data into three sets, i.e. generate an additional validation set which is used for hyperparameter tuning. As data acquisition is cumbersome with our measurement setup (cf. Chapter 4), we aim to have as much data for training and testing as possible. Therefore, we rely on the widely used alternative to a separate validation set, which is cross-validation (CV): If no validation set is available, the training set is split into N_{fold} equally sized subsets (*folds*). Each of these folds serves as a validation set for a model trained on the remaining $N_{\text{fold}} - 1$ folds, resulting in N_{fold} models. The reported accuracy for this hyperparameter selection is the average accuracy over the N_{fold} models. The number of folds in this work is set to $N_{\text{fold}} = 5$.

6.2.3 Hyperparameter Tuning

Depending on the selection of the classifier, various hyperparameters need to be selected to customize the model. We tune each model by an exhaustive search over an m-dimensional grid, where m is the number of hyperparameters we tune for this model. A larger and finer grid increases the required time for tuning, so tuning is limited to the most important hyperparameters. The model is trained and cross-validated as described above for each grid point, i.e. each possible combination of hyperparameter values within the selected limits. The hyperparameter combination with the highest accuracy is selected for the final model which is trained on the entire training dataset and then applied on the test data.

Our experiments have shown that for various subsets of the data (e.g. a smaller frequency range, fewer features, etc.), the result of the hyperparameter tuning remains mostly identical. Consequently, we use the hyperparameters specified for the classifiers in Section 6.3 for all evaluations. These hyperparameter choices are obtained from tuning the respective classifiers for data from all test subjects, all environments and measurements from all links. While this hyperparameter choice may not be optimal for every single subset of the data, the accuracy is comparable and more balanced between subsets of data in comparison to individually tuned models.

6.3 Classification based on Signal Energy

We begin with the simple case of signal energy / RSS measurements (raw data level 3), as the low number of low-complexity features (one scalar value per link) permits the use of simple "off-the-shelf" classifiers. In the following, we briefly introduce five common classification approaches for this type of data and select the most suitable for the posture classification task. For these descriptions, we follow the common notational convention of $\mathbf{x}_i$ for the N_{link}-dimensional feature vector of a sample i, and y_i as the corresponding class label (i.e. the posture or group).

For the problem at hand, we have $\mathbf{x}_i = [\widetilde{\text{RSS}}^{(p,q,1)}\ \widetilde{\text{RSS}}^{(p,q,2)}\ \ldots\ \widetilde{\text{RSS}}^{(p,q,N_{\text{link}})}]^T$ with the scaled scalar RSS values of the links, where every i is a snapshot measurement (p,q). In order to balance the absolute strength of the links and prevent dominance of the short-distance LOS links with high RSS, all values of the training set $\mathcal{S}_{\text{train}}$ are standardized via

$$\widetilde{\mathrm{RSS}}^{(p,q,l)} = \frac{\mathrm{RSS}^{(p,q,l)} - \mu_{\mathrm{RSS}^{(\cdot,\cdot,l)}}}{\sigma_{\mathrm{RSS}^{(\cdot,\cdot,l)}}} \tag{6.1}$$

for $\mathrm{RSS}^{(p,q,l)} \in \mathcal{S}_{\mathrm{train}}$. The mean $\mu_{\mathrm{RSS}^{(\cdot,\cdot,l)}}$ and standard deviation $\sigma_{\mathrm{RSS}^{(\cdot,\cdot,l)}}$ for each link l are computed across all snapshots from all postures in the training set. The same scaling (with mean $\mu_{\mathrm{RSS}^{(\cdot,\cdot,l)}}$ and standard deviation $\sigma_{\mathrm{RSS}^{(\cdot,\cdot,l)}}$ of the training data) is applied to the test data.

6.3.1 Support Vector Machines

SVMs are linear classifiers, i.e. in the feature space $\mathbf{x}$, the decision boundary is a hyperplane $\mathbf{wx} - b = 0$, where the binary decision between two classes is based on $\mathrm{sign}\,(\mathbf{wx} - b)$. The decision boundary is fitted in a way that maximizes the margins between the boundary and the closest data points ("support vectors"). For data which is not linearly separable, an adaptation of the loss function with a regularization hyperparameter C in the form

$$f_{\mathbf{w},b,C}(\mathbf{x}) = C\,\|\mathbf{w}\|^2 + \frac{1}{N_{\mathrm{link}}} \sum_{l=1}^{N_{\mathrm{link}}} \max\,(0, 1 - y_i(\mathbf{wx} - b)) \tag{6.2}$$

is applied, where C determines the tradeoff between exact decisions (small C) and generalization to avoid overfitting (large C). Different *kernel functions* allow a transformation of linearly inseparable data to a higher-dimensional space where linear separation is possible (the so-called *kernel trick*). The decision boundary and the determining support vectors can then be transformed back to the original domain, allowing for nonlinear decision boundaries in the original feature space. A common choice for such a kernel function is Gaussian (radial basis function (RBF)) with the standard deviation γ as a hyperparameter. The extension to multi-class classification is typically implemented in a one-versus-rest (OVR, also one-versus-all, OVA) or one-versus-one (OVO) fashion. The former may be imbalanced as the underlying decision ("class X" vs "not class X") can be biased due to data imbalance. The latter is more balanced for a similar number of training samples in all classes, but requires creating more binary classifiers and thus a longer training time, especially for a larger number of classes.

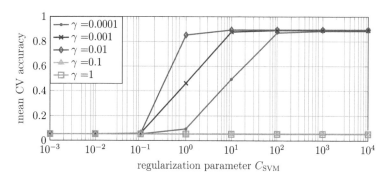

Figure 6.2.: CV accuracy for different hyperparameter combinations for SVM

Hyperparameters used in this work:

For the posture classification problem at hand, Fig. 6.2 shows the mean accuracy on the holdout part of the cross-validation dataset for various combinations of the most important hyperparameters. For clarity, only results for the RBF kernel are plotted, as SVMs with a linear kernel show a consistently weak performance. This indicates that the underlying data is not linearly separable in the original feature space. From Fig. 6.2 we observe that different combinations of the hyperparameters C and γ achieve a comparable cross-validation accuracy of $\approx 90\,\%$, e.g. $(C = 100, \ \gamma = 10^{-3})$ and $(C = 10, \gamma = 0.01)$.

For the posture classification task, we choose an OVO approach with RBF kernel and equal weight for all data points. As a result of the hyperparameter tuning, we set $C_{\mathrm{SVM}} = 10$ and the standard deviation of the RBF as $\gamma = 0.01$.

6.3.2 Logistic Regression

Despite its name, logistic regression (LR) is a classification method. Being a linear model similar to the SVM, it fits a hyperplane $\mathbf{w}\mathbf{x} = b$ as a decision boundary in the feature space $\mathbf{x}$. The additional smoothing of the boundary with the standard logistic function (also called sigmoid function) $f(x) = \frac{1}{1+e^{-x}}$ gives the method its name and is implemented as

$$f_{\mathbf{w},b}(\mathbf{x}) = \frac{1}{1 + e^{-(\mathbf{w}\mathbf{x}+b)}}. \tag{6.3}$$

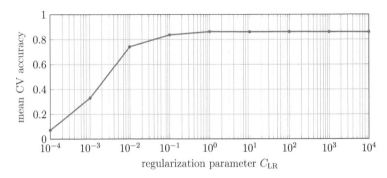

Figure 6.3.: CV accuracy for different choices of C_{LR}

in the likelihood function. In the multi-class extension provided in the library [95], a stochastic gradient descent solver is used to find the decision boundaries.

Hyperparameters used in this work:

Fig. 6.3 shows the influence of the L2-regularization hyperparameter C (analogously to the SVM). The mean cross-validation accuracy increases with stronger regularization up to $C = 1$. Consequently, we set $C_{\mathrm{LR}} = 1$ for the LR models used in this work.

6.3.3 *k*-Nearest Neighbors

Unlike other classification approaches like SVM or LR, kNN methods do not learn fixed decision boundaries. Instead, kNN is instance-based, i.e. it compares unseen test data points with the training data samples. Contrary to model-based approaches which store the decision boundaries and discard the training data after learning, kNN stores the training data for the classification: An unknown test sample is assigned the class of the majority of its k neighbors, which are determined as the closest training data points according to a distance metric. The key hyperparameters are the (typically odd) number of considered neighbors k, the distance metric, and the weighting scheme for the training samples.

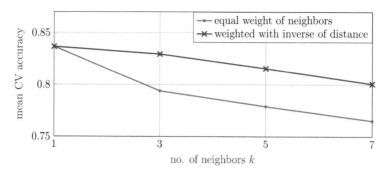

Figure 6.4.: CV accuracy for different hyperparameter combinations for kNN

Hyperparameters used in this work:

Fig. 6.4 illustrates the effect of different choices for the number of neighbors k and weighting schemes. For the posture classification task, considering fewer neighbors appears more promising. This is due to the fact that samples of different postures are often similarly close as samples from the same posture. Considering fewer neighbors increases the classification accuracy in this case, but may be more sensitive towards variations of the same posture. Weighting the neighbors inversely to their distance to the test sample increases the accuracy, as in most cases the closest neighbor is from the same posture (as we can see for $k = 1$). Based on the tuning results, we consider $k = 3$ neighbors weighted with the inverse of the Euclidean distance between them and the test sample. This choice has shown to be more accurate and robust than the nearest-neighbor-approach with $k = 1$ and a customized distance metric in our preliminary studies on the posture recognition problem [8, 36].

6.3.4   Random Forests

Every Random Forest (RF) is an ensemble model, i.e. a compilation of multiple weak models which are combined [97]. In the case of RFs, the sub-models are decision trees, each of which is based on a random selection of samples from the training dataset (usually with replacement, called "bagging"). For every split in the decision tree learning, only a random subset of all features is considered to avoid strong correlation between the trees of the ensemble. The majority vote of the predictions from all decision trees in the forest is then chosen as the prediction of the RF. The combination of multi-

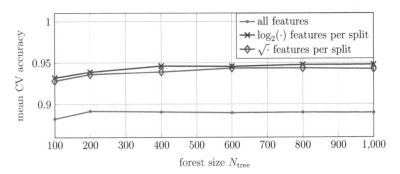

Figure 6.5.: CV accuracy for different hyperparameter combinations for RF

ple models and the random sample and feature selections increase the robustness and utility of RFs, which makes them a popular choice for a wide variety of classification tasks in various fields of application. The size of the forest (i.e. the number of decision trees constituting the RF) and the number of features per split in the trees are the most crucial hyperparameters [95]. An advantage of RFs lies in the possibility of parallelization, as the sub-models can be trained independently of each other.

Hyperparameters used in this work:

From Fig. 6.5 it is apparent that RFs can recognize the postures very reliably in the hold-out subsets during CV: An average CV accuracy above 0.88 across all hyperparameter combinations makes RFs a promising choice for the task at hand. Training more than $400 - 600$ trees in the forest does not result in further accuracy improvement. The number of features considered for each split at a tree node has a more significant influence: For the displayed case, considering $\lfloor \log_2 (153) \rfloor = 7$ features per split achieves a slightly higher average CV accuracy than $\lfloor \sqrt{153} \rfloor = 12$ features, but the difference is negligible. Using a subset of the available features at each split consistently proves to be better than considering all RSS values. In case of fewer features (see Section 7.3), we continue to use $\lfloor \log_2 (N_{\text{feat}}) \rfloor$ with N_{feat} as the number of links considered in the respective topology.

The RFs in the following are ensembles of 600 decision trees using $\lfloor \log_2 (N_{\text{feat}}) \rfloor$ features at every split (where N_{feat} is the total number of features available).

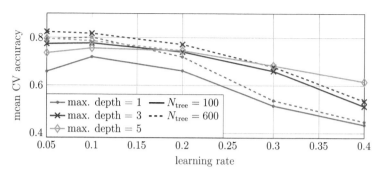

Figure 6.6.: CV accuracy for different hyperparameter combinations for GB

6.3.5   Gradient Boosting

Similarly to RFs, Gradient Boosting (GB) [98] is an ensemble learning method, i.e. multiple weaker learners are combined into one more powerful model. Decision trees are added to the ensemble iteratively, with each tree trained to refine the ensemble model based on the gradient of the likelihood function of the previous iteration (called "boosting"). Important hyperparameters are the learning rate α (i.e. the contribution of each tree), the number of decision trees (i.e. the number of iterations for model refinement) and the tree depth d, which should be limited to avoid overfitting. [46] The training process for GB takes significantly longer than the other methods due to the iterative procedure.

Hyperparameters used in this work:

Fig. 6.6 illustrates the cross-validation accuracy for different combinations of the tree depth, the number of estimators, and the learning rate. We observe that lower learning rates are generally better, but in turn take a longer time to learn the model. As expected, the accuracy increases with the number of models in the ensemble, i.e. with the number of iterations for refinement of the model. However, this effect saturates, so that the accuracy improvement for $N_{\text{tree}} > 600$ becomes negligible. For clarity of illustration, the respective plots are not shown in Fig. 6.6. A maximum tree depth $d = 3$ shows to be a good choice.

In the following, we choose a learning rate of $\alpha = 0.1$, a maximum tree depth of $d = 3$ and an ensemble size of $N = 600$ trees. All other hyperparameters are left with

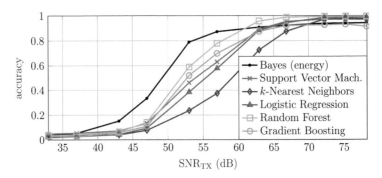

Figure 6.7.: Comparison of selected classifiers

their default values [95].

6.3.6 Comparison for Classifier Selection

In order to pick a suitable classifier for the RSS-based posture classification task, we compare the performance of the previously introduced and individually tuned classifiers on a test dataset comprising samples of all postures from all test subjects and environments. Fig. 6.7 illustrates the classification accuracy for all classifiers depending on the SNR. The classifiers are trained and tested individually for each SNR level to emulate a variation of the node transmit power, which would apply to both the calibration phase (training) as well as the time during operation (test).

We observe that for high SNR, the achieved accuracy exceeds 0.9. The maximum likelihood-inspired Bayes classifier and Gradient Boosting achieve a slightly lower accuracy than the other machine learning approaches for high SNR, but the maximum likelihood-inspired method is more robust towards the lower SNR range. Random Forests are a good balance between highest accuracy for high SNR and robustness for lower SNR.

Hence, we focus on RFs as the classifier of choice for signal energy-based posture classification in the remainder of this work. Note that while it may not be the optimal choice for every specific case covered, it shows a very satisfying accuracy overall, and is deemed suitable for solving the classification problem well enough to use it as a basis for a wearable posture recognition and fall prevention system.

6.3.7 Comparison with Maximum Likelihood Approach for Simplified System Model

In Chapter 5, we have introduced the maximum likelihood approach for the signal energy-based classification. The underlying simplified signal model does not consider differences between the posture snapshots due to variations of the postures and is thus of secondary importance to the design process of the envisioned posture recognition system. Nevertheless, we will take a brief excursion to compare the presented classifiers with the maximum likelihood approach based on the simplified signal model. For that purpose, we train all classifiers on all measurement snapshots, and thus deliberately skip the train-test-split in this case to ensure that the maximum likelihood classifier serves as an upper bound for the accuracy. It is important to note that this comparison has a more academic character and is not part of the application-focused system design process.

Fig. 6.8 illustrates the classification accuracy of the presented classifiers versus their theoretically optimal benchmark, the maximum likelihood classifier. We observe the

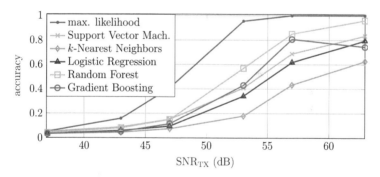

Figure 6.8.: Comparison of presented classifiers with maximum likelihood classifier

same relative performance differences as in Fig. 6.7 with a train/test split, with Random Forests outperforming the other machine learning approaches. Possible reasons for the observed difference to the maximum likelihood classifier as the upper bound for the lower SNR regime are a shortage of training data to accurately learn the noise distribution, as well as a potentially suboptimal choice of hyperparameters for this specific task. However, the accuracy of the Random Forest classifier approaches the upper bound for high SNR. This shows that with the selected hyperparameters, RFs can approximate the noncentral χ^2-distribution of the added noise most accurately.

6.4 Classification based on Sampled CIRs

CIR measurements as raw data offer a wide variety of features which can be used for classification. It is worthwhile to distinguish between simple scalar features (such as the delay of the first MPC, the maximum amplitude, or the easily obtainable signal energy as discussed in the previous section) and data sequences, for which the order of measurements (e.g. time samples of a CIR) is important. The order of the measured complex amplitude (or its magnitude) samples cannot be altered without changing the characteristics of the CIR. Consequently, they must not be considered individually but as a sequence, with the potential for additional characteristics beyond scalar values, e.g. the shape of a pulse which is stretched over multiple samples. On the contrary, the high dimensionality of the data requires more storage, processing power and more powerful classifiers. In the following, we will briefly introduce and evaluate two examples from the wide variety of available tools in the machine learning space: Dynamic Time Warping (DTW) as a distance metric for kNN, and convolutional neural networks (CNNs).

6.4.1 Dynamic Time Warping as Distance Metric for kNN

DTW [99] has been popular in speech processing for a long time, but it has proven to be a suitable tool for alignment of time series data of any kind [100]. Alignment is done by selectively repeating samples of either of the two sequences to be aligned, matching each sample of one sequence with its best corresponding sample (i.e. with the lowest distance) of the other sequence within defined limits. Fig. 6.9 illustrates this matching process: In Fig. 6.9a, the matched time samples between the upper sequence $X[i]$ and lower sequence $Y[j]$ are connected with yellow lines for illustration purposes, e.g. samples of the first plateau of $Y[j]$ are all matched to the first single sample of $X[i]$. This is seen in Fig. 6.9b in the horizontal part of the warping path, which is an illustration of the matching matrix: Time instance x_i in row $i \in [1, |X|]$ is matched to time instance y_j in column $j \in [1, |Y|]$ where the warping path is shown. Repeated samples can be seen as horizontal and vertical sections in the warping path. We further observe that most samples of y_j are delayed with respect to their respective matched counterparts x_j. While the direct Euclidean distance between the sequence vectors is large due to the delay, DTW accounts for the slight time shift and detects the similarity between the sequences which is apparent to the human observer. The delay is visible as the offset from the diagonal in the warping path in Fig. 6.9b. Several approaches

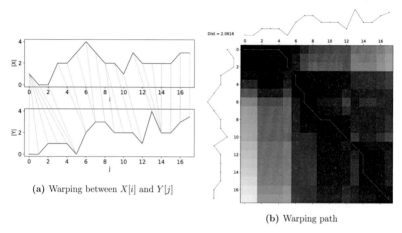

(a) Warping between $X[i]$ and $Y[j]$

(b) Warping path

Figure 6.9.: DTW principle (figure inspired by [101])

have been developed to increase the speed of sequence alignment with DTW, e.g. in [101–103]. The warping path is commonly restricted to speed up computation, e.g. by a limitiation of the maximum deviation from the diagonal [104], or by specifying a maximum number of repetitions for a single sample. Monotonicity and continuity of the warping path ensure that characteristics of the sequences are not repeated or omitted, respectively.

Using DTW as a distance metric between sequences allows for a classification with methods designed for scalar features. Especially kNN with a single neighbor ("1NN") and DTW as a distance metric has become popular due to its explainability and strong performance [105]. For our evaluations, we use the `dtaidistance` [106] and `tslearn` [107] libraries for Python.

We cannot assume perfect time synchronization for a realistic implementation low-cost, low-complexity wearable posture recognition system. Thus, the raw CIR measurements obtained with the VNA need to be adjusted: We add an individually random time shift sampled from a uniform distribution in the interval $\tau_{\text{shift}} \in [0, 10\,\text{ns}]$ to every CIR to emulate a respective timing uncertainty of the receiver. The Sakoe-Chiba band, i.e. the maximum deviation of the warping path from the diagonal, is set to the number of samples in 12 ns (the sum of the maximum possible random shift and 2 ns to account for posture variations). All links are individually standardized (across all CIRs for the respective link) to ensure equal weighting between strong and weak links.

Fig. 6.10 shows the classification accuracy for a 1NN-approach with DTW as distance metric. In the high SNR regime, the accuracy reaches 0.95. The comparatively rapid

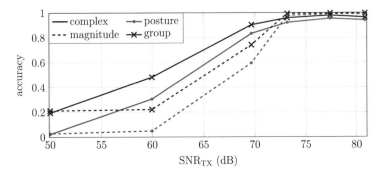

Figure 6.10.: Accuracy of 1NN-classifier with DTW distance metric

accuracy decrease with lower SNR, however, limits the approach to the high SNR regime. Compared to the signal energy-based kNN in Fig. 6.7, we observe that the magnitude-based DTW-1NN is less robust towards additional noise than its signal energy-based equivalent. With the random time shift and few MPCs, the additional information of the time-sampled magnitude of the CIR is mostly limited to the relative delay between the MPCs, as the absolute delay information is unknown. With very few MPCs, the signal energy captures most of the relevant information to distinguish the postures, as we can observe in the classification accuracies. The DTW distance metric, however, appears more sensitive to the purely noisy parts of the CIR without any significant signal MPCs, leading to a faster accuracy decline for a decreasing SNR.

The additional phase information significantly improves the classification accuracy for lower SNR, but the complex-valued case suffers from the same shortcomings as the magnitude-based approach, namely a low number of MPCs and a short duration of the relevant part of the CIR. Therefore, 1NN-classification with DTW is limited to the high SNR regime for our application, where it proves to be very accurate for both raw data levels.

6.4.2 Convolutional Neural Networks

Deep learning has become omnipresent in various signal processing domains. From the variety of available neural network architectures, CNNs are a useful means especially for image recognition and analysis. Inspired by the approach in [11] where CNNs are used for magneto-inductive posture recognition, we will explore the use of CNNs for posture recognition from CIRs in this section.

Detailed explanations of CNN architectures can be found in the literature. We will thus only briefly summarize the necessary components and their functionality. Each hidden layer of the CNN consists of multiple nodes with individual filters, which are convolved with the input image (hence the name). The filter size (i.e. its number of pixels in x and y) is smaller than the input image, so that the filter patch only covers a portion of the input image and is moved across it. This movement step size of the sliding window between the computation of two convolutions is referrred to as "stride". Depending on the stride, the output matrix is larger (small steps) or smaller (large steps). In order to properly examine the edges of an image, it is common to pad the image with zeros (analog to the discretization of the convolution of time signals from $-\infty$ to ∞). The filter coefficients for the convolution are learnt by the model over multiple epochs of backpropagation. The filter layer is typically followed by a pooling layer, which works in a similar way. However, instead of a convolution with optimized filter coefficients, the maximum value ("max pooling") or the average value ("average pooling") of the pixels in each patch are selected. This significantly reduces the data dimensionality by shrinking the image and focuses on the relevant areas of the input image. We use the more popular max pooling approach in the following. The output images of a pooling layer serve as an input volume (i.e. collection of images) for the following convolutional layer. All nodes of the following layer process all images of the previous layer, i.e. the network is fully connected. Fig. 6.11 illustrates the working principle of a convolution layer and a max pooling layer. In each neuron, the result is fed into a nonlinear activation function (similar to neuron activation in the human brain) before passing it on to the next layer (not shown in Fig. 6.11). Rectified Linear Units (ReLU) are a common choice as activation function, whose output equals $y_{\text{ReLU}} = \max(0, x)$ for the input x.

After sufficient reduction of the dimensionality, the image matrix of the last pooling layer is flattened and used as input for a fully connected multilayer perceptron (feed-forward neural network), whose outputs determine the classification result according

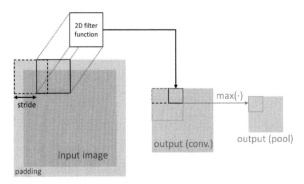

Figure 6.11.: CNN principle for one hidden layer

to the softmax function, i.e. selecting the output with the highest weight where the weights sum up to 1.

In order to use CNNs as described above, the data needs to be formatted as images. This is achieved by stacking the vectors of each link as rows of a matrix, which allows to store a posture snapshot as an $(N_{\text{link}} \times N_{\text{samp}})$-sized image. The brightness encodes the amplitude.

We observed that using frequency domain vectors instead of the CIRs in time domain for each link results in higher accuracy. The timing uncertainty changes the characteristic of the sparse CIRs, as the random time shift moves the bright pixels (i.e. the MPCs) around randomly in each CIR image row. A transformation into the frequency domain converts the time shift into a phase modification, which is eliminated by taking the absolute value of the frequency domain data, thus effectively removing the time shift from the data.

We further found that using the magnitude of the frequency domain data as input data for the CNN results in the highest accuracy for both complex CIR measurements (level 1 raw data) and their magnitude (level 2 raw data). In the following, the preprocessing is done as follows:

1. The measured CIR (or its magnitude, respectively) is randomly shifted in time ($\tau_{\text{shift}} \in [0, 10\,\text{ns}]$) to account for synchronization uncertainty as described above.

2. Artificial AWGN is added to emulate lower transmit power.

3. A time window of $T_{\text{win}} = 40\,\text{ns}$ is applied.

4. The CIR is transformed into the frequency domain using a fast Fourier transform.

5. The magnitude of the resulting complex-valued frequency response of each link is standardized using the mean and standard deviation for this link across all snapshots for all postures in the training set.

6. The standardized vectors for all links of a snapshot are stacked to form an image of $N_{\text{link}} \times N_{\text{samp}}$ pixels.

7. The resulting snapshot measurement image is used as input for the CNN.

The neural network is trained over 40 epochs with a learning rate of 0.001. We use (3×3) filters with ReLU activation and pad the image to maintain a constant size (padding of length 2 at all edges). The stride is set to $(1, 1)$.

Fig. 6.12 shows the accuracy for the CNN-based classification for both complex-valued CIRs and their magnitude as raw data.

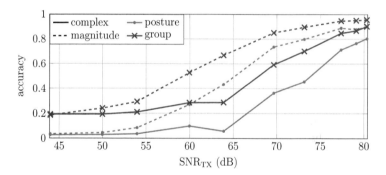

Figure 6.12.: Accuracy of posture classification with CNNs

Overall, the selected CNN configuration shows a comparable performance as the kNN approach with DTW distances, with the magnitude of the CIRs as the more robust raw data selection. These evaluations have not exploited the full potential of neural networks for the posture classification. However, the satisfying performance of the signal energy-based RF classifier in combination with its simple raw data measurement motivate us to explore this low-complexity option further and omit a deeper analysis of the more complex classifiers based on the sampled CIRs.

6.5 Summary

In this chapter, we have formulated the posture classification task as a machine learning problem, and evaluated the classification performance for a variety of common classifiers. A suitable set of hyperparameters was selected for each classifier. For the task at hand, signal energy-based Random Forests have shown to be a promising choice with an accuracy above 95 % for high SNR. Consequently, the potential performance advantage of more sophisticated models is limited. We further analyzed a 1NN-classifier with a DTW distance metric, as well as CNNs classifying based on the CIRs as image data. As neither of the two approaches outperform the significantly simpler Random Forests, we decide in favor of the latter for the following system parameter evaluations.

7

Robustness Analysis of Preferred Classifier-Feature-Combination

This chapter takes the next step towards the implementation of a wearable system by specifying a suitable parameter space for operation, including a reduction of the topology. We evaluate the robustness of our selection towards adversarial conditions like mismatches between training and test data regarding the test person, environment, and posture selection.

7.1 Frequency Range

The operating frequency range is a crucial parameter for a wearable wireless system. In Chapter 5, we have studied the influence of the frequency range for the simplified system model. In the following, we will briefly examine the dependency of the system performance for the RF classifier selected in Chapter 6 depending on the signal bandwidth and center frequency.

7.1.1 Bandwidth

For an analysis of the bandwidth, we determine the classification accuracy for (individually trained) Random Forest classifiers. The starting frequency of the selected band is the lower edge of the UWB range (3.1 GHz) in all cases. Fig. 7.1 illustrates the bandwidth dependency for the classification of postures and groups with $SNR_{TX} = 80\,dB$ for $B = 5.4\,GHz$. We observe an increase in accuracy with increasing bandwidth for both levels of classification granularity. Note that as pointed out in Section 4.2.2, the additional measurement energy from the additional frequency bins increases the SNR

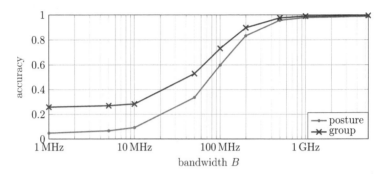

Figure 7.1.: Influence of bandwidth

proportionately to the bandwidth. Analogously, the power spectral density limit for UWB allows for a transmit power proportionate to the occupied bandwidth.

In order to separate the influence of the bandwidth and the SNR, we briefly examine the posture classification accuracy while keeping the SNR constant, as shown in Fig. 7.2. For higher SNR and corresponding high accuracy, the influence of the bandwidth is negligible. For lower SNR, the low accuracy is hardly influenced by the bandwidth. For higher SNR, the accuracy improves by 17 percentage points from 0.31 to 0.48 with increasing bandwidth over one decade (100 MHz to 1 GHz). Due to the constant SNR, the difference is most likely due to the frequency diversity, which results in more stable signal energy measurements for the varying channels around the body. As the accuracy is sensitive to small changes in the measured energy (cf. the steeper part of the curves in Fig. 6.7), stable measurements provide a strong advantage in this regime. In the following, we use a bandwidth of $B = 1\,\text{GHz}$.

7.1.2   Center Frequency

We will further briefly examine the influence of the center frequency of the WBAN signals within the available UWB domain. With the signal bandwidth fixed to $B = 1\,\text{GHz}$, we vary the center frequency in the interval between $f_{c,\text{min}} = 3.6\,\text{GHz}$ and $f_{c,\text{max}} = 8\,\text{GHz}$. The results are illustrated in Fig. 7.3. The achieved accuracy is higher in the lower frequency range, with negligible differences below $f_c = 6\,\text{GHz}$. Consequently, we use the lower UWB band starting from $f_{\text{min}} = 3.1\,\text{GHz}$ in the following evaluations.

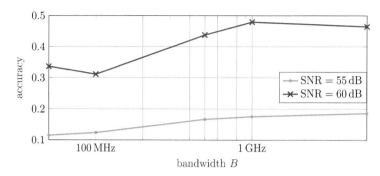

Figure 7.2.: Influence of bandwidth for constant SNR

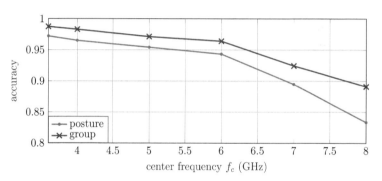

Figure 7.3.: Influence of center frequency

7.1.3 Frequency Offset

The IEEE standards governing body area networks with UWB state clear requirements regarding the maximum tolerated center frequency offset. According to the
IEEE 802.15.4 standard [64], the maximum permitted offset is up to ±40 ppm for some
implementations, and 0.6 MHz for DSSS. Its extension IEEE 802.15.4z [108], a carrier
offset tolerance of ±20 ppm is specified, which equals a maximum offset between transmitter and receiver oscillator of 40 ppm≙144 kHz for our selected center frequency of
3.6 GHz. Using only the signal energy of the CIR, any time-dependent phase variation
due to a frequency mismatch between transmitter and receiver is discarded. Furthermore, the short CIRs result in the frequency response changing slowly over frequency.
Hence, a frequency offset. i.e. a slight shift of the applied bandpass filter window,

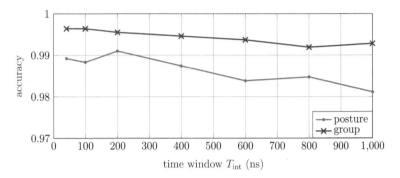

Figure 7.4.: Influence of integration window

would not result in a major change of the received signal energy over the large bandwidth. We thus conclude that the system is robust towards frequency offsets by design.

7.2   Time Window

This section has been published in a similar form in [37].

The integration window T_{int} is a key parameter for the implementation of the UWB IR receiver. Choosing the integration window for the energy measurement longer than the delay spread (typically $\tau_{\text{RMS}} \approx 30\,\text{ns}$ in our case) leads to a lower SNR due to additional collected noise. Thus, a lower accuracy for a longer integration time can be expected. Fig. 7.4 illustrates the influence of the integration window length T_{int} on the accuracy.

We observe that for all topologies the high SNR is sufficient even for longer integration times, as even for $T_{\text{int}} = 1\,\mu\text{s}$ the accuracy is above 0.98. For the selected transmit power, the SNR reserve is large enough so that a longer integration window does not reduce the accuracy significantly. Note that the SNR is indirectly proportionate to the length of the integration window. This is in line with Fig. 6.7, as a time window of $1\,\mu\text{s}$ is equivalent to a SNR decrease by $10\,\text{dB}$ compared to a window of $100\,\text{ns}$ (as used for the evaluation in Fig. 6.7): The accuracy decrease from $\text{SNR}_{\text{TX}} = 78\,\text{dB}$ to $\text{SNR}_{\text{TX}} = 68\,\text{dB}$ is negligible.

7.3 WBAN Topology Minimization

Note: A similar version of this section has been published in [37].

Reducing the number of nodes in a WBAN has various advantages such as a reduction of the overall power demand, which is of utmost importance for wearable battery-operated systems. Furthermore, fewer nodes improve comfort for the user and reduce costs and effort of integrating the system into clothing, which makes it comfortably wearable.

A topology reduction implies choosing a subset of the 153 links whose signal energies (i.e. features) help distinguishing the postures. The existing literature provides a wide variety of established feature selection methods, e.g. in [109–112]. Kohavi and John [113] elaborate on feature relevance and the optimality of subsets, and give examples for wrapper and filter approaches, which do and do not take the classifying algorithm into account, respectively.

From the wide variety of available feature selection methods, we choose a wrapper approach based on the permutation importance. On a separate validation dataset, the effect of randomly permuting each individual feature across all training measurements on the classification accuracy is determined. A strong decrease in accuracy indicates high feature importance and vice versa [97]. This allows to create a ranking between the links which, however, is classifier-dependent, i.e. it may differ for an SVM and an RF. We determine the feature importance for an RF classifier using training and validation data for all postures, subjects and environments.

The 10 most important links are illustrated in Fig. 7.5a. The single most important links are (i) between both ankles, (ii) between the right ankle and the right hip, and (iii) between the right knee and the right chest. This is plausible as their characteristics change with (i) the relative position of the legs (which is especially important for a distinction of the walking snapshots), and (ii & iii) the relative position of torso and legs, respectively. A ranking of nodes is established by ordering all nodes according to the sum of their respective link importances, as shown in Fig. 7.5b. Nodes on the shoulders and upper arms are of lower importance, which is understandable as their relative position to the other nodes does not vary as much as on the extremities. Note that the link importance depends on the postures considered, e.g. postures differing solely with regard to the arm position require links to/from arm-mounted nodes to be distinguished (cf. also Section 5.3.1). However, we consider the large variety of our underlying postures sufficiently representative for this analysis.

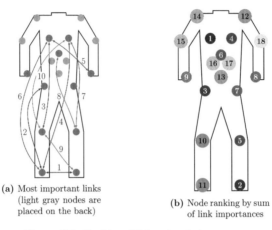

(a) Most important links
(light gray nodes are
placed on the back)

(b) Node ranking by sum
of link importances

Figure 7.5.: Ranking of link and node importance

Links and nodes can be selected according to various schemes. In the following, we will distinguish 3 *architectures*: First, we use only the most important links, denoted as "SL" for *Selected Links*. For the second type of architecture, we assume that all nodes can send and receive (denoted as "TRX") and consider a fully connected set of the most important nodes. Note that this type of topology requires transceiver hardware (but not full duplex operation) at all nodes, which may be more power-demanding than the third architecture considered, which assumes dedicated nodes for sending and receiving (denoted "TX–RX"). These nodes with reduced capabilities (either TX or RX) reduce the node implementation complexity compared to TRX nodes at the cost of the number of available links for a fixed number of nodes.

7.3.1 Selected Links

The most intuitive and straightforward approach to pick features for the classification problem at hand is to use the signal energy of the most important N_{link} links, i.e. the most important features. However, this may result in a large number of required nodes and does not necessarily result in a simplification of the topology, as nodes which are only required for a single selected link need to be deployed. In practice, selecting only the top 10 or top 20 links results in 12 and 16 required nodes, respectively. Note that other links between the nodes are not considered if their rank is lower than the top N_{link}. We therefore consider this case primarily for comparison.

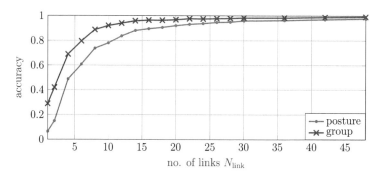

Figure 7.6.: Classification accuracy using the most important links

Fig. 7.6 shows the classification accuracy depending on the number of links N_{link}, which are selected by decreasing importance regardless of the nodes involved. We observe that the accuracy of both posture and group classification rapidly increase with the number of links used. Using only the 14 most important links for classification already yields an accuracy above 0.9, which gradually increases to 0.98 with up to 34 additional links. The slow increase indicates that the relevant information to distinguish the respective classes is contained in the most important 20 links.

7.3.2 Best TRX Nodes

As purely selecting links based on their importance results in topologies with an undesirably large number of nodes, we consider selecting nodes instead of links to control the density of the topology. This architecture is based on the ranking of nodes based on the sum of their respective link importances as shown in Fig. 7.5b. The most important N_{node} nodes form a fully connected WBAN, in which all links are considered. This topology results in $\frac{N_{node} \cdot (N_{node}-1)}{2}$ links, i.e. the number of links scales quadratically with N_{node}.

Fig. 7.7a shows the classification accuracy depending on the number of nodes. We observe a rapid increase in accuracy when increasing the number of nodes up to 6. Additional nodes have a minor effect, and no significant increase of the accuracy can be achieved with more than 10 nodes.

Fig. 7.7b compares the accuracy of the TRX architecture to the previously introduced SL architecture based on the number of links in the topology. The quadratic increase

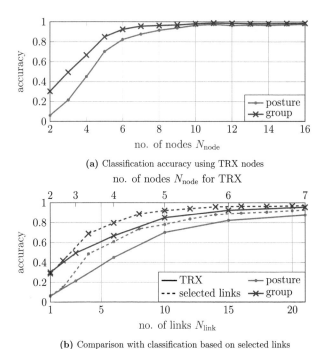

(a) Classification accuracy using TRX nodes

(b) Comparison with classification based on selected links

Figure 7.7.: Classification accuracy of TRX architecture

in the number of links with an increasing number of TRX nodes is illustrated by the upper second x-axis. The SL architecture consistently outperforms the TRX topologies with the same number of links, which is expected. However, the number of required nodes (and thus the density of the topology) differs significantly: In order to achieve a posture classification accuracy of 0.8, 6 TRX nodes are required. The same accuracy can be achieved with 12 selected links, which, however, span across 12 different nodes (not shown in plot).

7.3.3 Hierarchical Topology

As an alternative to the dense topologies of the SL architecture and the requirement of the TRX architecture for nodes with transmit- and receive-capability, we consider a third architecture with a hierarchy between the dedicated roles of the nodes, namely agents (TX) and anchors (RX).

Based on the ranking in Fig. 7.5b, the N_{anc} most important nodes are selected as anchors and the following N_{ag} nodes as agents. We assume no communication between the anchors for this topology. Consequently, only the $(N_{\mathrm{ag}} \cdot N_{\mathrm{anc}})$ links between agents and anchors are considered. In the following, we use the notation "$N_{\mathrm{ag}} \times N_{\mathrm{anc}}$".

The advantage of this distinction lies in the possible simplification of the hardware implementation of the agent nodes: Transmitting agents do not require receiver circuitry, so their functionality can be limited to a periodic broadcast of a unique identifier. Anchors require slightly more complexity, as they conduct the energy measurement and serve as relays forwarding the pre-processed and aggregated information to a central processing unit (e.g. a smartphone). The concept for this architecture has also been discussed in [8] and Chapter 3, and is covered in more detail in Chapter 8.

Fig. 7.8 shows the classification accuracy for various agent-anchor combinations for both granularity levels of classification (posture and group). As only links between agents and anchors are used, the sparse topologies with very few nodes perform poorly. On the other hand, the densest topologies achieve an accuracy of 0.95 even for posture classification. Topologies with the same number of links tend to perform similarly (e.g. 4×4 and 8×2), which gives great flexibility to the system designer. For instance, consider a scenario where a group classification accuracy of ≥ 0.9 is required. This can be achieved with either a 4×4 or a 10×1 topology. If the particular implementation of the anchors is significantly more complex and power-hungry than the agents, the 10×1 topology is the better choice. If the power demand and complexity of anchors and agents is comparable, the 4×4 topology is preferred as it requires only 8 nodes in total (instead of 11 for the 10×1 topology). Consequently, the best topology choice depends on the application and hardware architecture of the nodes and can be selected from the variety of options based on the performance comparison in Fig. 7.8.

In order to put the hierarchical topologies into perspective, we briefly compare them with the TRX architecture: Consider the 14×2 topology and 8 fully connected TRX nodes with 28 links each. While the performance is comparable (posture classification accuracy ≈ 0.9) due to the same number of links, the hierarchical topology requires twice as many nodes as the fully connected network. In turn, these can have a simpler architecture. The ratio of agents and anchors plays a significant role as well: The accuracy of the best hierarchical topology with 8 nodes (4×4) is almost as high as the accuracy with 8 fully connected nodes (0.85 vs. 0.88).

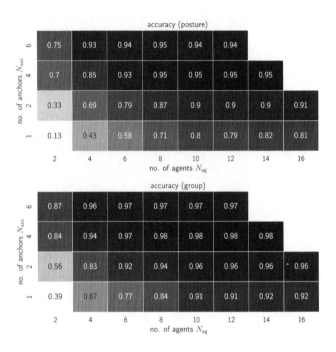

Figure 7.8.: Classification accuracy for hierarchical topologies

7.3.4 Topology Selection

In the following, we will focus on three selected topologies in more detail. Thereby we omit the practically infeasible architecture with a focus on selected links. We pick the simplest topologies from the TRX- and the hierarchical architecture with a posture classification accuracy ≥ 0.9, i.e. (i) 8 TRX nodes, as well as (ii) 6×4 and (iii) 10×2 TX–RX nodes. We consider two hierarchical topologies due to the potential differences in node complexity as outlined above. Fig. 7.9 illustrates the choices.

7.4 Environment

For ubiquitous posture recognition, consistently high classification accuracy regardless of the surroundings is desired. However, obtaining reliable training data for all possible

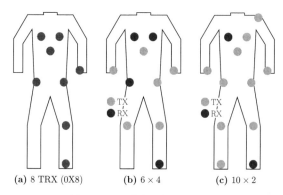

(a) 8 TRX (0X8) **(b)** 6 × 4 **(c)** 10 × 2

Figure 7.9.: Selected topologies

environments of operation is not possible. Nevertheless, we aim to gain some insight into the robustness towards different surroundings of the user in this section. We do so by comparing the classification accuracy for different training/test data combinations, which differ in terms of the surroundings. As explained in Chapter 4, the collected data contains measurements from both *open* indoor environments (i.e. without objects in the immediate surroundings of the test subject) and *cluttered* spaces (i.e. with a variety of reflecting and scattering objects nearby). Despite the different clutter constellations for different test subjects, we only distinguish between open and cluttered surroundings here. It is noteworthy that this harsh contrast represents a worst-case scenario, as most common environments in private homes are neither fully open nor extremely cluttered. However, for a robustness evaluation the worst-case scenario is suitable, as it outlines the sensitivity of the system. We distinguish two cases, namely matched and mismatched environments. Therefore, training and test datasets each comprise only measurements from a single environment. Fig. 7.10 illustrates the classification accuracy for matched and mismatched cases, each averaged over both possible train-test combinations (matched: open-open, clutter-clutter / mismatched: open-clutter, clutter-open).

For matching train and test environments, the accuracy is consistently high. However, for a mismatch in training and test environment, the posture classification accuracy drops by about 50 percentage points across all three considered topologies. The decrease in group classification accuracy is less drastic with about 25 percentage points, with the hierarchical topologies performing marginally better.

While the underlying test case reflects a worst-case scenario as outlined previously,

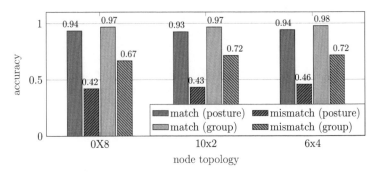

Figure 7.10.: Classification accuracy depending on the environment

these results underline the requirement to collect training data for various different environments during the calibration process. The increased robustness from a mixed training set (i.e. training data from both environments) can be observed from the the the previously presented results for mixed training and test sets in Fig. 7.7 and 7.8: With posture recognition accuracies of 0.91 (8 TRX), 0.93 (6 × 4), and 0.90 (10 × 2), the respective results for mixed training and testing lie in a similar range as for matched environments. Including a variety of surroundings in the training data therefore increases the robustness of the system significantly.

7.5   Cross-Subject

A key promise in the field of e-health is personalized healthcare tailored to the needs of the individual patient. Nevertheless, an evaluation of the sensitivity of the posture recognition system towards the individual user is beneficial, as it provides insight in the calibration requirements: Does the system need an individual training for each user or does it work reliably with a fixed out-of-the-box training set? Analogous to the analysis for different environments, we aim to develop an intuition for the system's robustness across different test persons. While the data from our three subjects of different physique is suitable to provide basic insights, the available data does not suffice for a desirable universal calibration. For this evaluation, we use the measurements from all environments. We distinguish two different data splits: Individual calibration, i.e. data from the same person for training and test, and cross-subject, i.e. using data from two subjects for training and the third one for testing. The results shown in Fig. 7.11 are averaged over all three training data pairings (P1 & P2, P1 & P3, P2 & P3).

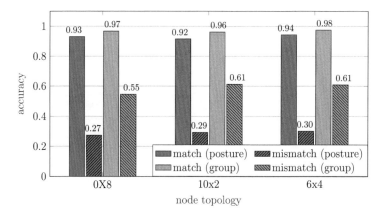

Figure 7.11.: Classification accuracy depending on the test person

While the accuracy for both posture and group classification is strong for matching training and test subjects, it drops significantly (by more than 60 and 30 percentage points, respectively) in case of a mismatch. Thus, the system is highly sensitive towards differences between the individual subjects. As the same behavior is observed for the selected case of training with P1 and P3 while testing with P2, we can conclude that training with data from a short and a tall person does not improve the performance for a person of average height for testing. The different link distances from varying proportions of the physique would thus require either a large training dataset consisting of a variety of subjects and/or more advanced data augmentation techniques to use training data from one person for another test subject. Data augmentation is a very active field of current research, a brief overview can be found e.g. in [114].

Overall, the differences are significantly larger than for the environmental change. This is understandable due to the large differences in the test subjects' physique, and deviations in the exact placement of the on-body antennas, which directly affect all links. Changes to the environment do not affect all links simultaneously, but only those which observe additional multipath components in the presence of obstacles. In the existing literature (e.g. [89]), the influence of the human body on WBAN channels is also observed to exceed the influence of the surrounding environment. For the system at hand, the dependency on the person outlines the requirement for a separate calibration (i.e. training phase) for each individual subject, or a respective representation of the physique in the training data: When the respective physique is part of a mixed training

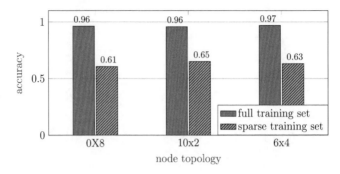

Figure 7.12.: Group classification accuracy for sparse data

set (as in Fig. 7.7 and 7.8), accuracies ≥ 0.9 are achieved with the selected topologies. Hence, a sufficiently diverse training set ensures robustness.

7.6 Sparse Data

An inherent problem of classification tasks is their requirement for adequate and comprehensive training data, whose patterns are matched with the test data patterns. The wide variety of possible human body postures can never be fully captured to form a comprehensive training set: It is impossible to foresee all possible postures and include them in the training data. Moreover, the training effort would be infeasibly tremendous. As a consequence, new data in the field will always include unknown postures. Classifying these unknown postures as their correct group, however, is possible and essential: Especially for fall prevention, an imminent fall and its direction needs to be detected correctly to initiate appropriate countermeasures. In order to test the robustness of group classification towards unknown postures, we randomly split the postures of each group into two equally sized sets and use one for training and the other one for testing. For an odd number of postures in a specific group, the remaining posture is also assigned to the training set. Consequently, training and test set consist of entirely different postures.

Fig. 7.12 shows the group classification accuracy for the different topologies, averaged over five different random splits into training/test postures. For comparison we include the performance of the respective classifiers with access to samples from all postures in the training data.

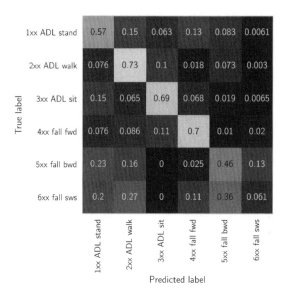

Figure 7.13.: Confusion matrix for testing with unknown postures (8 TRX topology)

All topologies exhibit similar behavior. For a test set with completely unknown postures, the group classification accuracy drops by $30 - 35\%$ compared to the case with all test postures in the training data. Many postures exhibit sufficient similarities with other postures from different groups, so that depending on the train/test split, misclassifications occur. For the case of sparse training data, the confusion matrix for classification on group level is shown in Fig. 7.13 with the classification result on the x-axis and the true group of the test data on the y-axis. The comparatively higher accuracy for walking, sitting and forward falling postures shows that postures from these groups are significantly more similar to each other than to postures from other groups. This is understandable, as the walking postures only show major differences in the relative leg orientation (but are otherwise very similar to each other), sitting postures have similar and distinct LOS/NLOS situation for the leg–torso links, and in most falling forward postures the arms are extended forward. We further see that falling backwards and sideways are difficult to identify. Postures from both groups are comparatively often misclassified as standing or walking, and thus the imminent fall is not detected correctly. While this is the worst type of error in the fall prevention setting, it needs to be pointed out that these groups are underrepresented in the training data in comparison to the other groups, with only three (backwards) and one (sideways) posture(s) in the sparse training set. While the exact results differ slightly for the other two topology candidates, the large trends (such as a higher accuracy for walking, sitting and falling forward) can also be observed for them.

7.7 Summary

In this chapter we have selected an operation point in the frequency domain at $f_c = 3.6\,\text{GHz}$ with $1\,\text{GHz}$ bandwidth for our posture recognition system, and chosen a window length of $T_{\text{int}} = 100\,\text{ns}$ for the energy detector receiver. Three candidates for reliable topologies were identified in a systematic way, which were consequently examined for their robustness regarding shortcomings of the training data set. This analysis has outlined the requirement for accurate training data for all three topology candidates, including different environments and individual training for each user.

8

Outline of Preferred System Implementation

In this chapter, we will use the previous findings to choose a set of parameters to provide a suggestion for a system implementation with a suitable topology. While the comprehensive details of the full system implementation are beyond this work, we will briefly revisit the WBAN architecture introduced in Chapter 3 and illustrate the options for implementation on a conceptual level. A simulative analysis of the expected performance of the proposed system within the limits of the available data concludes the chapter.

8.1 Summary of Previous Findings

The evaluations in the previous chapters have shown that a wearable posture recognition system based on UWB is a feasible endeavour. Even with simple signal energy measurements, a high classification accuracy can be achieved for a posture set with partially very similar postures. Therefore, there is limited benefit from more complex and cumbersome time sequence measurements. From a variety of common machine learning approaches, Random Forests have emerged as a suitable choice for a signal energy-based classifier. We have shown that 1 GHz bandwidth in the lower UWB range is a suitable frequency range. This choice allows for a relaxation of the measurement time synchronization, as the duration of the measurement time hardly affects the classification accuracy. As long as all relevant MPCs of the CIR ($\tau_{\mathrm{rms}} \leq 30\,\mathrm{ns}$) are within the measurement window, the energy measurement is sufficiently accurate. The time window for the measurement can be chosen as long as $1\,\mu\mathrm{s}$ without compromising on the accuracy (assuming no interference between nodes).

We have selected three topology candidates in a systematic way based on the link importances: A topology with 8 transceiver nodes, as well as two hierarchical topologies with 6 agents and 4 anchors, and 10 agents and 2 anchors, respectively. We observe that all three topology candidates perform similarly in case of various shortcomings of the underlying data or mismatches between training and test data, such as differences in the measurement surroundings, different test subjects, or different subsets of postures. In the following, we will elaborate on the topology candidates further in order to make a suitable selection.

8.2 Implementation Concepts

Note: Parts of this chapter were published in [8] in a similar fashion.

We favor a hierarchical topology over a WBAN with 8 transceivers, as the number of additional nodes in case of a hierarchical topology (2 for 6×4, 4 for 10×2) is outweighed by the flexibility and simplicity of nodes.

We hence follow the hierarchical WBAN concept introduced in Chapter 3, consisting of low-complexity agents broadcasting their UIDs to body-mounted anchor relays, which measure the received signal energy and forward the information to a (potentially off-body) central unit. This separation of functionality allows for simpler node architectures and centralizes the information required for posture recognition at the central unit. It further grants flexibility regarding the physical layer implementation of both hops. In the following, we will briefly discuss two possible concepts, using DSSS and a combination of UWB IR and BLE, respectively.

8.2.1 DSSS

As mentioned in Section 3.4.2, the agent-individual spreading codes for DSSS serve as UIDs. The ultra-low complexity agents transmit their UIDs $c_a(t)$ modulated onto a carrier $\cos(\omega_a t)$ as illustrated in Fig. 8.1. Agents with additional sensors could transmit sensor data as payload with their UID as an identifying spreading code.

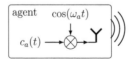

Figure 8.1.: DSSS agent node

Using DSSS for both hops allows for a very simple relay design, such as the example based on frequency division duplexing (FDD) depicted in Fig. 8.2. The "amplify-mix-and-forward" relay includes a multiplication with a relay identifier waveform $c_r(t)$, which enables the assignment of the measurement to the respective first hop link at the processing unit. FDD is achieved via the upconversion with the difference $\Delta\omega$ of

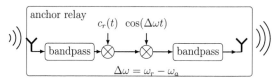

Figure 8.2.: DSSS anchor relay with FDD

the carrier frequencies of the first (ω_a) and second hop (ω_r). Bandpass filters prevent self-interference from the relay output to its input. FDD for the prevention of self-interference allows for continuous operation and is simple to implement. Time division duplexing (TDD) requires time synchronization between agents and anchors and long guard intervals due to varying distances of the agent-anchor links.

As a trade-off for the simple node designs, the complexity of this approach is shifted to the receiver at the central unit. To identify and assign the received signals from the anchors, the receiver needs to correlate the received signal with shifted copies of the joint agent-anchor sequences, i.e. the possible product sequences of UIDs $c_a(t - \tau_a)$ and $c_r(t - \tau_r)$. This requires a high computational effort and parallelization, due to the large variety of possible relative shifts between the agent- and anchor-UIDs, i.e. combinations of τ_a and τ_r. These vary with the level of synchronization between the on-body nodes, as well as with the times of flight of the first hop, i.e. with the posture. The receiver of the central unit is thus significantly more complex than the nodes of the lower layers. For nodes integrated in clothing, partial synchronization can be established via a wired connection across all nodes within the same piece of garment, which allows to reduce the receiver complexity: The relative offset between agents and anchors integrated in the same piece of clothing is known up to few nanoseconds, as only the times of flight remain unknown for these nodes.

8.2.2   Impulse Radio & Bluetooth LE

An attractive alternative to the correlation-based receiver in the DSSS approach is a change of the physical layer technology at the relay. With a smartphone as the central processing unit, BLE is an attractive choice, as it is supported out of the box and well established in a wide variety of handhelds, whereas UWB chips are still limited to flagship smartphones. Fig. 8.3 shows the block diagram of a respective UWB IR agent. A change from UWB for the first hop to BLE for the second hop

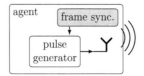

Figure 8.3.: UWB IR agent node

increases the complexity of the relays, but allows using UWB IR for the first hop. The combination of UWB IR and BLE requires a conversion unit for decoding, processing and encoding, which increases power consumption at the relay. Fig. 8.4 illustrates the respective relay architecture. With this approach, agents and anchors need to be

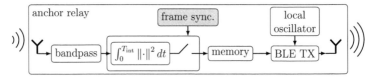

Figure 8.4.: Anchor relay with UWB IR receiver and BLE transmitter

frame-synchronized to avoid interference between agents. Each agent uses a dedicated time slot for the pulse transmission. The identification of the agent then just happens via the allocated time-slot. With integration windows and thus frames of $\geq 100\,\text{ns}$, the synchronization does not need to be in the sub-nanosecond range, as outlined in more detail in Section 8.2.3. Frame synchronization can be implemented using wired connections (for nodes integrated in garments) or wirelessly.

Another advantage of this more complex anchor design is the possibility for preprocessing: Measuring the signal energy of the first hop with an energy detector reduces the amount of data to forward to a single scalar value per link. Memory at the anchor allows to accumulate measurements from multiple agents before forwarding them jointly. The required memory size, the system's update rate and the number of BLE transmissions need to be carefully balanced.

Advantages of this approach are the low duty cycle and the simplicity of reception due to UWB IR, as simple energy detectors at the anchor relays avoid the need for excessive correlation at the central unit and allow to completely detach the first and second hop. Due to the resulting increased complexity of the relays, choosing a topology with a strong imbalance between the number of agents and anchors is beneficial. From the previously selected topology candidates, the 10×2 topology is thus favored over the 6×4 topology. The comparable performance (cf. Chapter 7) allows for as few anchors as possible without compromising on classification accuracy. The on-body placement of the agents and anchors is displayed in Fig. 8.5.

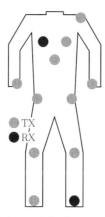

Figure 8.5.: Node placement for the proposed 10×2 topology

8.2.3 Timing

Interference between agents disturbs the energy measurements at the anchors. The separate clocks at all 12 nodes therefore require a periodic time synchronization to compensate their individual clock drifts. While developing a complete synchronization scheme is beyond the scope of this thesis, it is worthwhile to examine the requirements for a sufficiently high update rate of the system.

With an integration window of $T_{\text{int}} = 100\,\text{ns}$ for each of the agents, the complete measurement time for a single snapshot measurement cycle for all links takes $N_{\text{ag}}T_{\text{int}} = 1\,\mu\text{s}$, as the anchor relays measure simultaneously. With an additional guard interval of $T_{\text{guard}} = 100\,\text{ns}$ between the anchor measurement windows, the snapshot measurement time is $N_{\text{ag}}(T_{\text{int}} + T_{\text{guard}}) \approx 2\,\mu\text{s}$.

Regarding the timing synchronization between the nodes, we assume initial time synchronicity at $t = t_0$. An initial synchronization such that the received CIR is centered in the integration window at the anchor relay ensures the highest robustness towards clock drifts. With a delay spread of $\tau_{rms} \approx 30$ ns, a timing offset of $\tau_{off} \geq (T_{int} - \tau_{rms})/2 \approx 35$ ns causes the CIR to partially fade out of the integration window, potentially leading to an erroneous energy measurement. The individual clock-generating quartz crystals at each node are assumed to have a stability of ± 50 ppm. Thus, in the worst case a time difference between two nodes of $\Delta \tau = 35$ ns occurs after an operating time of $\frac{\Delta \tau}{2 \cdot 50 \cdot 10^{-6}} = 0.35$ ms after synchronization (≈ 175 snapshot measurements in case of continuous operation). Consequently, the synchronization process needs to be repeated accordingly to ensure accurate energy measurements.

With short measurement cycles of less than $2\,\mu s$ per snapshot (including guard times), the limiting factor for the update rate is the BLE connection to the central unit. For BLE advertising, the fixed transmission interval can be set as low as 20 ms with an additional random delay of up to 10 ms [51], which therefore limits the update rate of the proposed system to 30 ms. The reaction time of human reflexes, which naturally prevent falls e.g. when tripping, lies in the same range. Each BLE packet has a maximum payload of 251 bytes, so that every packet can contain up to 25 complete snapshot measurements of 10 links to each relay sampled with an 8-bit-ADC. This allows to reduce the duty cycle to a single measurement per 30 ms to save energy. Alternatively, instead of a reduction of the measurement frequency, the system's robustness can be increased in two ways: Averaging the measured energy values over multiple measurements before using them as features for the posture classification increases stability. This averaging could be performed at the anchor relay already. Alternatively, identifying the posture for every measurement and taking a majority vote across the classification results can prevent misclassifications that occur as single outliers.

8.3 Simulative Verification of the Preferred System

Finally, we will simulatively examine the expectable performance of the envisioned system. Thereby, we evaluate the posture classification accuracy for a combination of the parameters selected in Chapter 7. The SNR is kept as a variable parameter, as it is one of the easiest system parameters to adapt and of significant importance for a wearable and battery-powered system.

We use training and test data from all subjects and all environments. The SNR is modified using additive white Gaussian noise as described in Chapter 5. Furthermore, a random time shift (i.i.d. randomly selected from a uniform distribution over $[0, 20]$ ns is applied to every CIR.

As in the previous evaluations, we emulate lower pulse energies by scaling the additive noise accordingly. Fig. 8.6 illustrates the performance for the proposed system operating with $B = 1$ GHz bandwidth around $f_c = 3.6$ GHz, an integration window of $T_{\text{int}} = 100$ ns, and a hierarchical 10×2 topology. A posture classification accuracy of

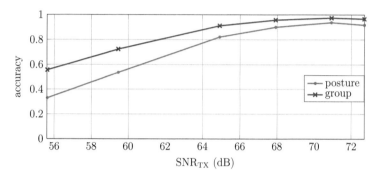

Figure 8.6.: Classification accuracy of proposed implementation

0.9 can be achieved using single pulse measurements with a transmit SNR ≈ 68 dB. Consequently, the selected setup with a simple 10×2 topology and UWB IR as a physical layer for the first hop allows for reliable, low-power posture recognition with a single pulse measurement per link.

9

Conclusion & Outlook

In this thesis, a system concept for wearable wireless posture recognition based on WBAN signaling has been proposed. We approached this task from a measurement-based perspective due to the infeasibility to accurately model the challenging and constantly varying wireless channel conditions around the human body. A comprehensive measurement campaign comprising 43 postures from six groups laid the foundation for the data-based approach towards posture recognition. With measurements from three test subjects of different physique in open and cluttered environments, we consider the set of measurements sufficiently diverse for the task. The precise complex-valued CIR measurements with a Vector Network Analyzer have enabled us to selectively discard information and evaluate posture classification on three different levels of available raw data. In a final system realization, the choice of the physical layer and the implementation of the wireless on-body nodes determines the level of available raw data.

A first feasibility evaluation based on a simplified measurement-based signal model indicated the feasibility of the posture recognition task at hand. Consequently, we used the diverse data to determine suitable machine learning models for the posture classification. From the vast variety of available classifiers, we compared selected candidates. For time series data, both a nearest-neighbor-classifier with Dynamic Time Warping as a distance metric as well as convolutional neural networks proved to be suitable for the high SNR range. In both cases, using the magnitude of the measured CIRs has shown to be more robust than the complex-valued samples. For posture classification based on the signal energy of each link, Random Forests are the classifier of choice. With comparable peak performance (above 95% accuracy) and lower complexity, we have selected signal energy-based Random Forests for further evaluation.

A bandwidth of 1 GHz in the lower UWB range has proven to be suitable for the energy measurement. The regulatory limits for UWB transmission permit an implementation with UWB IR which achieves a sufficiently high SNR with a single pulse.

We further systematically selected the WBAN topology, i.e. reduced the number of on-body nodes while choosing a suitable placement for the remaining antennas. For our system proposal, we have selected an asymmetric topology with ten low-complexity transmitting agents and two receiving anchor relays. These anchor relays measure the received signal energy of the agent signals and forward the aggregated data via BLE to the user's smartphone, which acts as a central unit and performs the computation-intensive posture classification.

Such a system is well suited to perform reliable recognition of the body posture for the diverse set of measurements in this work. Furthermore, it can be implemented with low-complexity and lightweight on-body nodes. It is, however, limited in its ability to cope with mismatches in training and test data: Our analysis has shown the importance of diverse training data from different environments, and individual training data for each user. Extending the data basis towards additional environments and especially older test subjects, as well as exploring the possibilites of data augmentation to increase the robustness are the recommended next step for future work.

Looking forward, the system is predestined for combinations with other solutions, which can increase the robustness of posture recognition or complement and extend the system. An obvious example is a combination with inertial measurement units (IMUs) with accelerometers and gyroscopes. While IMU-based posture recognition is challenging for static postures, it can be used to support the wireless signaling-based approach. Another synergy is the transmission of IMU data: As the IMU measurements need to be aggregated to identify the posture, the wireless data transmission from the IMUs at the agents to aggregating anchors can be utilized in our proposed way to increase the confidence in the classification result. Moreover, a combination with fitness trackers or smartwatches is promising: Activity recognition by a fitness tracker can be improved by a time series of detected postures, e.g. a series of walking postures. In return, sensors monitoring vital functions such as the heart rate can be used to dynamically adapt the update rate of the posture recognition: A rising heart rate indicates potentially fall-prone activity such as standing up from a bed or chair, and thus an increased risk of falling. In such cases, the posture can be monitored more closely, while a lower duty cycle during resting activities saves battery. This further opens up the possibility of continuous semi-supervised or unsupervised learning to refine the model and adapt it to individual habits of the user.

Appendix

A.1 Maximum Likelihood Classifier for Complex CIR

A coherent receiver directly measures the superposition of complex CIR and the complex noise components. The measured noise is therefore complex Gaussian distributed, resulting in a likelihood function as a mixture of Gaussian components:

$$\mathcal{L}_p = \frac{1}{Q_p \sqrt{(2\pi)^{N_{\mathrm{link}} \cdot N_{\mathrm{samp}}}} \, \sigma^{N_{\mathrm{link}} \cdot N_{\mathrm{samp}}}} \sum_{q=1}^{Q_p} \exp\left(-\frac{\left\|\mathbf{m} - \mathbf{s}^{(p,q)}\right\|^2}{\sigma^2}\right). \qquad (A.1)$$

For the maximum likelihood estimate

$$\hat{p} = \arg \max_p \mathcal{L}_p = \arg \max_p \frac{1}{Q_p} \sum_{q=1}^{Q_p} \exp\left(-\frac{\left\|\mathbf{m} - \mathbf{s}^{(p,q)}\right\|^2}{\sigma^2}\right). \qquad (A.2)$$

the leading factor can be omitted. Note that the sum over Q_p different snapshots for each posture prevents the use of the log-likelihood in this case.

A.2 Maximum Likelihood Classifier for Magnitude of CIR

As outlined in Chapter 3, information about the signal phase may not be available depending on the hardware implementation of the receiving nodes. The measured magnitude of the CIRs is Rician distributed with the posture snapshots $\mathbf{s}^{(p,q)}$ as respective means.

We obtain the resulting likelihood for posture p as

$$\mathcal{L}_p = \frac{1}{Q_p} \sum_{q=1}^{Q_p} \prod_{k=1}^{N_{\text{link}} \cdot N_{\text{samp}}} \frac{|m_k|}{\sigma^2/2} \exp\left(\frac{-(|m_k|^2 + |s_k^{(p,q)}|^2)}{\sigma^2}\right) I_0\left(\frac{|m_k|\,|s_k^{(p,q)}|}{\sigma^2/2}\right) \quad \text{(A.3)}$$

with $I_0(\cdot)$ denoting the modified Bessel function of the first kind with order zero, and the maximum likelihood estimate

$$\hat{p} = \arg\max_p \frac{1}{Q_p} \sum_{q=1}^{Q_p} \prod_{k=1}^{N_{\text{link}} \cdot N_{\text{samp}}} \frac{|m_k|}{\sigma^2/2} \exp\left(\frac{-(|m_k|^2 + |s_k^{(p,q)}|^2)}{\sigma^2}\right) I_0\left(\frac{|m_k|\,|s_k^{(p,q)}|}{\sigma^2/2}\right)$$

$$\text{(A.4)}$$

$$= \arg\max_p \frac{1}{Q_p} \sum_{q=1}^{Q_p} \underbrace{\exp\left(-\frac{\sum_{k=1}^{N_{\text{link}} N_{\text{samp}}} |s_k^{(p,q)}|^2}{\sigma^2}\right)}_{\gamma^{(p,q)}} \prod_{k=1}^{N_{\text{link}} \cdot N_{\text{samp}}} \exp\left(\frac{-|m_k|^2}{\sigma^2}\right) I_0\left(\frac{|m_k|\,|s_k^{(p,q)}|}{\sigma^2/2}\right)$$

$$\text{(A.5)}$$

$$= \arg\max_p \frac{1}{Q_p} \sum_{q=1}^{Q_p} \gamma^{(p,q)} \prod_{k=1}^{N_{\text{link}} \cdot N_{\text{samp}}} I_0\left(\frac{|m_k|\,|s_k^{(p,q)}|}{\sigma^2/2}\right), \quad \text{(A.6)}$$

where the factor $\gamma^{(p,q)}$ is independent of the measurement and can thus be precomputed. The terms which do not depend on the posture are omitted for the maximum likelihood decision.

A.3 Maximum Likelihood Classifier for Signal Strength

In this section, the received signal strength for each link is measured separately, the respective measurement for the CIR $\mathbf{h}^{(p,q,l)}$ of a link is denoted as $\mathbf{m}^{(\cdot,\cdot,l)} \in \mathbb{C}^{N_{\text{samp}} \times 1}$, so that in agreement with (5.1) we have

$$\mathbf{m}^{(\cdot,\cdot,l)} = \mathbf{h}^{(p,q,l)} + \mathbf{n}^{(\cdot,\cdot,l)} \quad \text{(A.7)}$$

For the the maximum likelihood estimate for a measurement of the received signal strength per link

$$\mathrm{RSS}_{\mathbf{m}}^{(\cdot,\cdot,l)} = \left\| \mathbf{m}^{(\cdot,\cdot,l)} \right\|^2 = \sum_{k=1}^{N_{\mathrm{samp}}} (m_{I,k}^{(\cdot,\cdot,l)})^2 + (m_{Q_p,k}^{(\cdot,\cdot,l)})^2 \tag{A.8}$$

$$\text{with} \quad m_I^{(\cdot,\cdot,l)} = \mathfrak{Re}\{\mathbf{m}^{(\cdot,\cdot,l)}\}, \quad m_{Q_p}^{(\cdot,\cdot,l)} = \mathfrak{Im}\{\mathbf{m}^{(\cdot,\cdot,l)}\}, \tag{A.9}$$

we make use of the fact that the elements of $\mathbf{m}_I^{(\cdot,\cdot,l)}$ and $\mathbf{m}_{Q_p}^{(\cdot,\cdot,l)}$ are i.i.d. Gaussian distributed. A sum of squared i.i.d. Gaussian random variables X_i with unit variance follows a χ^2-distribution. We define

$$\overline{\mathbf{m}^{(\cdot,\cdot,l)}} = \begin{bmatrix} \mathbf{m}_I^{(\cdot,\cdot,l)} \\ \mathbf{m}_{Q_p}^{(\cdot,\cdot,l)} \end{bmatrix} \tag{A.10}$$

so that

$$\overline{\mathbf{m}^{(\cdot,\cdot,l)}} \sim \mathcal{N}\left(\overline{\mathbf{h}^{(p,q,l)}} = \begin{bmatrix} \mathbf{h}_I^{(p,q,l)} \\ \mathbf{h}_{Q_p}^{(p,q,l)} \end{bmatrix}, \frac{\sigma^2}{2} \mathbf{I} \right) \tag{A.11}$$

and scale the variance to obtain

$$\frac{\sqrt{2}}{\sigma} \overline{\mathbf{m}^{(\cdot,\cdot,l)}} = \widetilde{\mathbf{m}^{(\cdot,\cdot,l)}} \sim \mathcal{N}\left(\frac{\sqrt{2}}{\sigma} \overline{\mathbf{h}^{(p,q,l)}}, \mathbf{I} \right). \tag{A.12}$$

Based on (A.8) we have

$$\mathrm{RSS}_{\mathbf{m}}^{(\cdot,\cdot,l)} = \frac{\sigma^2}{2} \sum_{k=1}^{2N_{\mathrm{samp}}} \left(\widetilde{m_k^{(l)}} \right)^2 = \frac{\sigma^2}{2} \left\| \widetilde{\mathbf{m}^{(\cdot,\cdot,l)}} \right\|^2 \sim (\text{scaled}) \; \chi^2 \left(2N_{\mathrm{samp}}, \lambda \right). \tag{A.13}$$

With the respective means from (A.12) we obtain the noncentrality parameter

$$\lambda^{(p,q,l)} = \sum_{k=1}^{N_{\mathrm{samp}}} \left(\frac{\sqrt{2}}{\sigma} h_{I,k}^{(p,q,l)} \right)^2 + \left(\frac{\sqrt{2}}{\sigma} h_{Q_p,k}^{(p,q,l)} \right)^2 \tag{A.14}$$

$$= \frac{2}{\sigma^2} \sum_{k=1}^{N_{\mathrm{samp}}} \left(h_{I,k}^{(p,q,l)} \right)^2 + \left(h_{Q_p,k}^{(p,q,l)} \right)^2 \tag{A.15}$$

$$= \frac{2}{\sigma^2} \left\| \mathbf{h}^{(p,q,l)} \right\|^2 = \frac{2}{\sigma^2} \mathrm{RSS}_{\mathbf{h}}^{(p,q,l)} \tag{A.16}$$

for the noncentral χ^2-distribution. Scaling accordingly via $f_Y(y) = \frac{2}{\sigma^2} f_X\left(\frac{2}{\sigma^2}x\right)$ provides the likelihood function

$$
\mathcal{L}_p^{(l)} = \frac{1}{Q_p} \sum_{q=1}^{Q_p} \frac{1}{\sigma^2} \exp\left(-\frac{\frac{2}{\sigma^2}\mathrm{RSS}_{\mathbf{m}}^{(\cdot,\cdot,l)} + \lambda^{(p,q,l)}}{2}\right) \left(\frac{2\,\mathrm{RSS}_{\mathbf{m}}^{(\cdot,\cdot,l)}}{\sigma^2\,\lambda^{(p,q,l)}}\right)^{(N_{\mathrm{samp}}-1)/2}
$$
$$
I_{N_{\mathrm{samp}}-1}\left(\sqrt{\lambda^{(p,q,l)}\frac{2}{\sigma^2}\mathrm{RSS}_{\mathbf{m}}^{(\cdot,\cdot,l)}}\right) \quad (A.17)
$$

for the link l of posture p.

The maximum likelihood estimate for the RSS case is the product of all link probabilities and therefore evaluates to

$$
\hat{p} = \arg\max_p \frac{1}{Q_p} \sum_{q=1}^{Q_p} \prod_{l=1}^{N_{\mathrm{link}}} \exp\left(-\frac{\mathrm{RSS}_{\mathbf{m}}^{(\cdot,\cdot,l)}}{\sigma^2} - \frac{\lambda^{(p,q,l)}}{2}\right) \left(\frac{2\,\mathrm{RSS}_{\mathbf{m}}^{(\cdot,\cdot,l)}}{\sigma^2\,\lambda^{(p,q,l)}}\right)^{(N_{\mathrm{samp}}-1)/2}
$$
$$
I_{N_{\mathrm{samp}}-1}\left(\frac{\sqrt{2\,\lambda^{(p,q,l)}\,\mathrm{RSS}_{\mathbf{m}}^{(\cdot,\cdot,l)}}}{\sigma}\right). \quad (A.18)
$$

With the noncentrality parameter from (A.16) we obtain

$$
\hat{p} = \arg\max_p \frac{1}{Q_p} \sum_{q=1}^{Q_p} \prod_{l=1}^{N_{\mathrm{link}}} \exp\left(-\frac{\mathrm{RSS}_{\mathbf{m}}^{(\cdot,\cdot,l)} + \mathrm{RSS}_{\mathbf{h}}^{(p,q,l)}}{\sigma^2}\right)
$$
$$
\left(\frac{\mathrm{RSS}_{\mathbf{m}}^{(\cdot,\cdot,l)}}{\mathrm{RSS}_{\mathbf{h}}^{(p,q,l)}}\right)^{(N_{\mathrm{samp}}-1)/2} I_{N_{\mathrm{samp}}-1}\left(\frac{2}{\sigma^2}\sqrt{\mathrm{RSS}_{\mathbf{h}}^{(p,q,l)}\,\mathrm{RSS}_{\mathbf{m}}^{(\cdot,\cdot,l)}}\right)
$$
$$
= \arg\max_p \frac{1}{Q_p} \sum_{q=1}^{Q_p} \exp\left(-\frac{\sum_{l=1}^{N_{\mathrm{link}}} \mathrm{RSS}_{\mathbf{h}}^{(p,q,l)}}{\sigma^2}\right)
$$
$$
\prod_{l=1}^{N_{\mathrm{link}}} \exp\left(-\frac{\mathrm{RSS}_{\mathbf{m}}^{(\cdot,\cdot,l)}}{\sigma^2}\right) \left(\frac{\mathrm{RSS}_{\mathbf{m}}^{(\cdot,\cdot,l)}}{\mathrm{RSS}_{\mathbf{h}}^{(p,q,l)}}\right)^{(N_{\mathrm{samp}}-1)/2} I_{N_{\mathrm{samp}}-1}\left(\frac{2}{\sigma^2}\sqrt{\mathrm{RSS}_{\mathbf{h}}^{(p,q,l)}\,\mathrm{RSS}_{\mathbf{m}}^{(\cdot,\cdot,l)}}\right).
$$
$$
(A.19)
$$

List of Variables & Notation

Symbol	Unit	Name
B	Hz	bandwidth
d	m	distance
E	Ws	energy
f	Hz	frequency
G	1	gain
i	A	electric current
N	1	count
p	-	posture
P	W	power
q	-	snapshot
Q_p	-	collection of snapshots of posture p
t	s	time
T	s	duration
v	V	voltage
Z	Ω	impedance
φ	°	azimuth angle
λ	m	wavelength
σ	1	standard deviation
θ	°	elevation angle
ω	s^{-1}	circular frequency

Notation	Description
x	scalar variable
$\mathbf{x}$	vector
x_k	k-th scalar entry of vector $\mathbf{x}$
$\mathbf{X}$	matrix
$(\cdot)^T$	transpose
$(\cdot)^H$	conjugate transpose
$\|\cdot\|$	$L2$-norm

List of Figures

List of Acronyms

ADC   analog-to-digital converter

ADL   activities of daily living

AWGN   additive white Gaussian noise

BLE   Bluetooth Low Energy

CDF   cumulative distribution function

CIR   channel impulse response

CNN   convolutional neural network

CV   cross-validation

DSSS   Direct Sequence Spread Spectrum

DTW   Dynamic Time Warping

ED   energy detector

FDD   frequency division duplexing

GB   Gradient Boosting

IDFT   inverse discrete Fourier transform

IF   intermediate frequency

IMU   inertial measurement unit

IR   impulse radio

kNN   k-nearest neighbors

LOS   line of sight

LR   logistic regression

MPC   multipath component

NLOS   non-line of sight

OVO   one-versus-one

PDF   probability density function

PN   pseudo-random noise

RBF   radial basis function

RF   Random Forest

RSS   received signal strength

RSSI   received signal strength indicator

SNR   signal-to-noise ratio

SVM   support vector machine

TDD   time division duplexing

tSNE   t-distributed Stochastic Neighbor Embedding

UID   unique identifier

UWB   ultra-wideband

VNA   Vector Network Analyzer

WBAN wireless body area network

Bibliography

[1] Z. Mekonnen, "Time of Arrival Based Infrastructureless Human Posture Capturing System," Ph.D. dissertation, ETH Zurich, 2016, Zurich, Switzerland.

[2] J. Bryan and Y. Kim, "Classification of human activities on UWB radar using a support vector machine," in *2010 IEEE Antennas and Propagation Society International Symposium*, 2010, pp. 1–4.

[3] Z. Baird, S. Rajan, and M. Bolic, "Classification of Human Posture from Radar Returns Using Ultra-Wideband Radar," in *2018 40th Annual International Conference of the IEEE Engineering in Medicine and Biology Society (EMBC)*, 2018, pp. 3268–3271.

[4] Z. Zhang, C. Conly, and V. Athitsos, "A Survey on Vision-Based Fall Detection," in *Proceedings of the 8th ACM International Conference on PErvasive Technologies Related to Assistive Environments*, ser. PETRA '15, 2015.

[5] M. Yu, A. Rhuma, S. M. Naqvi, L. Wang, and J. Chambers, "A Posture Recognition-Based Fall Detection System for Monitoring an Elderly Person in a Smart Home Environment," *IEEE Transactions on Information Technology in Biomedicine*, vol. 16, no. 6, pp. 1274–1286, 2012.

[6] D. Brulin, Y. Benezeth, and E. Courtial, "Posture Recognition Based on Fuzzy Logic for Home Monitoring of the Elderly," *IEEE Transactions on Information Technology in Biomedicine*, vol. 16, no. 5, pp. 974–982, 2012.

[7] J. He, H. Li, and J. Tan, "Real-time Daily Activity Classification with Wireless Sensor Networks using Hidden Markov Model," in *2007 29th Annual International Conference of the IEEE Engineering in Medicine and Biology Society*, 2007, pp. 3192–3195.

[8] R. Heyn and A. Wittneben, "Detection of Fall-Related Body Postures from WBAN Signals," in *GLOBECOM 2020 - 2020 IEEE Global Communications Conference*, 2020, pp. 1–6.

[9] H. Schulten, F. Wernli, and A. Wittneben, "Learning-Based Posture Detection Using Purely Passive Magneto-Inductive Tags," in *2021 IEEE Global Communications Conference (GLOBECOM)*, 2021, pp. 1–7.

[10] H. Schulten and A. Wittneben, "Experimental Study of Posture Detection Using Purely Passive Magneto-Inductive Tags," in *WCNC 2022 - IEEE Wireless Communications and Networking Conference*, 2022, pp. 1–6.

[11] ——, "Robust Multi-Frequency Posture Detection based on Purely Passive Magneto-Inductive Tags," in *ICC 2022 - 2022 IEEE International Conference on Communications (ICC)*, 2022, pp. 1–5.

[12] Y. Tang, Z. Peng, L. Ran, and C. Li, "iPrevent: A novel wearable radio frequency range detector for fall prevention," in *2016 IEEE International Symposium on Radio-Frequency Integration Technology (RFIT)*, 2016, pp. 1–3.

[13] Y. Tang, Z. Peng, and C. Li, "An experimental study on the feasibility of fall prevention using a wearable K-band FMCW radar," in *2017 United States National Committee of URSI National Radio Science Meeting (USNC-URSI NRSM)*, 2017, pp. 1–2.

[14] J. G. Argañarás, Y. T. Wong, G. Pendharkar, R. Z. Begg, and N. C. Karmakar, "Obstacle detection with mimo 60 ghz radar for fall prevention," in *2019 IEEE Asia-Pacific Microwave Conference (APMC)*, 2019, pp. 366–368.

[15] M. Pech, H. Sauzeon, T. Yebda, J. Benois-Pineau, and H. Amieva, "Falls Detection and Prevention Systems in Home Care for Older Adults: Myth or Reality?" *Journal of Medical Internet Research (JIMR) Aging*, vol. 4, no. 4, pp. 1–8, 2021.

[16] H. Hawley-Hague, E. Boulton, A. Hall, K. Pfeiffer, and C. Todd, "Older adults' perceptions of technologies aimed at falls prevention, detection or monitoring: a systematic review," *International Journal of Medical Informatics*, vol. 83, no. 6, pp. 416–426, 2014.

[17] Y. Delahoz and M. Labrador, "Survey on Fall Detection and Fall Prevention Using Wearable and External Sensors," *Sensors*, vol. 14, no. 10, p. 19806–19842, Oct 2014.

[18] X. Hu and X. Qu, "Pre-impact fall detection," *BioMedical Engineering OnLine*, vol. 15, no. 61, pp. 1–16, 2016.

[19] L. Ren and Y. Peng, "Research of Fall Detection and Fall Prevention Technologies: A Systematic Review," *IEEE Access*, vol. 7, 2019.

[20] C. Wang, W. Lu, M. Narayanan, S. Redmond, and N. Lovell, "Low-power Technologies for Wearable Telecare and Telehealth Systems: A Review," *Biomedical Engineering Letters (BMEL)*, vol. 5, no. 1, pp. 1–9, 2015.

[21] S. Sigg, M. Scholz, S. Shi, Y. Ji, and M. Beigl, "RF-Sensing of Activities from Non-Cooperative Subjects in Device-Free Recognition Systems Using Ambient and Local Signals," *IEEE Transactions on Mobile Computing*, vol. 13, no. 4, pp. 907–920, 2014.

[22] M. Huang, J. Liu, Y. Gu, Y. Zhang, F. Ren, X. Wang, and J. Li, "Your WiFi Knows You Fall: A Channel Data-Driven Device-Free Fall Sensing System," in *ICC 2019 - 2019 IEEE International Conference on Communications (ICC)*, 2019, pp. 1–6.

[23] D. C. Ranasinghe, R. L. S. Torres, A. P. Sample, J. R. Smith, K. Hill, and R. Visvanathan, "Towards falls prevention: A wearable wireless and battery-less sensing and automatic identification tag for real time monitoring of human movements," in *2012 Annual International Conference of the IEEE Engineering in Medicine and Biology Society*, 2012, pp. 6402–6405.

[24] T. Wilding, E. Leitinger, U. Muehlmann, and K. Witrisal, "Modeling Human Body Influence in UWB Channels," in *2020 IEEE 31st Annual International Symposium on Personal, Indoor and Mobile Radio Communications*, 2020, pp. 1–6.

[25] Q. H. Abbasi, A. Sani, A. Alomainy, and Y. Hao, "Arm movements effect on ultra wideband on-body propagation channels and radio systems," in *2009 Loughborough Antennas Propagation Conference*, 2009, pp. 261–264.

[26] T. Kumpuniemi, T. Tuovinen, M. Hämäläinen, K. Y. Yazdandoost, R. Vuohtoniemi, and J. Iinatti, "Measurement-based on-body path loss modelling for UWB WBAN communications," in *2013 7th International Symposium on Medical Information and Communication Technology (ISMICT)*, March 2013, pp. 233–237.

[27] T. Kumpuniemi, M. Hämäläinen, K. Y. Yazdandoost, and J. Iinatti, "Categorized UWB on-Body Radio Channel Modeling for WBANs," *Progress In Electromagnetics Research B*, vol. 67, pp. 1–16, 2016.

[28] X. Huang, F. Wang, J. Zhang, Z. Hu, and J. Jin, "A Posture Recognition Method Based on Indoor Positioning Technology," *Sensors*, vol. 19, no. 6, 2019.

[29] C. Einsmann, M. Quirk, B. Muzal, B. Venkatramani, T. Martin, and M. Jones, "Modeling a wearable full-body motion capture system," in *Ninth IEEE International Symposium on Wearable Computers (ISWC'05)*, 2005, pp. 144–151.

[30] E. Farella, Pieracci, D. Brunelli, L. Benini, B. Ricco, and A. Acquaviva, "Design and implementation of wimoca node for a body area wireless sensor network," in *Proceedings of the 2005 Systems Communications*, ser. ICW '05. USA: IEEE Computer Society, 2005, p. 342–347.

[31] E. Farella, A. Pieracci, L. Benini, and A. Acquaviva, "A Wireless Body Area Sensor Network for Posture Detection," in *11th IEEE Symposium on Computers and Communications (ISCC'06)*, 2006, pp. 454–459.

[32] M. Quwaider and S. Biswas, "Body Posture Identification Using Hidden Markov Model with a Wearable Sensor Network," in *Proceedings of the ICST 3rd International Conference on Body Area Networks*, no. 19, 2008.

[33] Y. Geng, J. Chen, R. Fu, G. Bao, and K. Pahlavan, "Enlighten Wearable Physiological Monitoring Systems: On-Body RF Characteristics Based Human Motion Classification Using a Support Vector Machine," *IEEE Transactions on Mobile Computing*, vol. 15, no. 3, pp. 656–671, March 2016.

[34] I. C. Paschalidis, W. Dai, D. Guo, Y. Lin, K. Li, and B. Li, "Posture Detection with Body Area Networks," in *Proceedings of the 6th International Conference on Body Area Networks*, 2011, pp. 27–33.

[35] X. Yang, M. Fang, A. Ren, Z. Zhang, Q. H. Abbasi, A. Alomainy, K. Mehran, and Y. Hao, "Reverse recognition of body postures using on-body radio channel characteristics," *IET Microwaves, Antennas & Propagation*, vol. 11, no. 9, pp. 1212–1217, 2017.

[36] R. Heyn and A. Wittneben, "Comprehensive Measurement-Based Evaluation of Posture Detection from Ultra Low Power UWB Signals," in *IEEE International Symposium on Personal, Indoor and Mobile Radio Communications (PIMRC)*, 2021, pp. 1–6.

[37] ——, "WBAN Node Topologies for Reliable Posture Detection from On-Body UWB RSS Measurements," in *ICC 2022 - IEEE International Conference on Communications*, 2022, pp. 5298–5303.

[38] R. Heyn, M. Kuhn, H. Schulten, G. Dumphart, J. Zwyssig, F. Trösch, and A. Wittneben, "User Tracking for Access Control with Bluetooth Low Energy," in *2019 IEEE 89th Vehicular Technology Conference (VTC2019-Spring)*, 2019, pp. 1–7.

[39] H. Schulten, M. Kuhn, R. Heyn, G. Dumphart, F. Trösch, and A. Wittneben, "On the Crucial Impact of Antennas and Diversity on BLE RSSI-Based Indoor Localization," in *2019 IEEE 89th Vehicular Technology Conference (VTC2019-Spring)*, 2019, pp. 1–6.

[40] G. Dumphart, R. Kramer, R. Heyn, M. Kuhn, and A. Wittneben, "Pairwise Node Localization From Differences in Their UWB Channels to Observer Nodes," *IEEE Transactions on Signal Processing*, vol. 70, pp. 1576–1592, 2022.

[41] G. Dumphart, M. Kuhn, A. Wittneben, and F. Trösch, "Inter-node distance estimation from multipath delay differences of channels to observer nodes," in *ICC 2019 - 2019 IEEE International Conference on Communications (ICC)*, 2019, pp. 1–7.

[42] S. Robinovitch, F. Feldman, Y. Yang, R. Schonnop, P. M. Leung, T. Sarraf, J. Sims-Gould, and M. Loughin, "Video capture of the circumstances of falls in elderly people residing in long-term care: an observational study," *The Lancet*, vol. 381, no. 9860, pp. 47 – 54, 2013.

[43] G. Dumphart, B. Bitachon, and A. Wittneben, "Magneto-Inductive Powering and Uplink of In-Body Microsensors: Feasibility and High-Density Effects," in *2019 IEEE Wireless Communications and Networking Conference (WCNC)*, 2019, pp. 1–6.

[44] G. Dumphart, "Magneto-Inductive Communication and Localization: Fundamental Limits with Arbitrary Node Arrangements," Ph.D. dissertation, ETH Zurich, 2020, Zurich, Switzerland.

[45] C. Bishop, *Pattern Recognition And Machine Learning*, 1st ed., ser. Information Science and Statistics. Springer, 2006.

[46] A. Burkov, *The Hundred-Page Machine Learning Book*, 2019.

[47] U. Mengali and A. D'Andrea, *Synchronization Techniques For Digital Receivers*, ser. Applications of Communications Theory, R. W. Lucky, A. S. Acampora, T. Li, and W. H. Tranter, Eds. Springer Science+Business Media, 1997.

[48] A. Goldsmith, *Wireless Communications*. Cambridge University Press, 2005.

[49] H. Meyr, M. Moeneclaey, and S. A. Fechtel, *Digital Communication Receivers: Synchronization, Channel Estimation, and Signal Processing*. Wiley, 1998.

[50] "IEEE Standard for Local and metropolitan area networks - Part 15.6: Wireless Body Area Networks," *IEEE Std 802.15.6-2012*, pp. 1–271, 2012.

[51] "Bluetooth Specification Revision v5.2," *Bluetooth Core Specifications*, vol. 0, pp. 1–3256, 2019.

[52] M. Schwartz, *Mobile Wireless Communications*. Cambridge University Press, 2004.

[53] C. S. Miller, "A technique for rapid detection of spread spectrum sequences," in *1997 IEEE Aerospace Conference*, vol. 1, 1997, pp. 83–90.

[54] R. Gold, "Optimal binary sequences for spread spectrum multiplexing (Corresp.)," *IEEE Transactions on Information Theory*, vol. 13, no. 4, pp. 619–621, 1967.

[55] ——, "Maximal recursive sequences with 3-valued recursive cross-correlation functions (Corresp.)," *IEEE Transactions on Information Theory*, vol. 14, no. 1, pp. 154–156, 1968.

[56] E. Leitinger, P. Meissner, C. Rüdisser, G. Dumphart, and K. Witrisal, "Evaluation of position-related information in multipath components for indoor positioning," *IEEE Journal on Selected Areas in Communications*, vol. 33, no. 11, pp. 2313–2328, 2015.

[57] J. Möderl, F. Pernkopf, and K. Witrisal, "Car Occupancy Detection Using Ultra-Wideband Radar," in *2021 18th European Radar Conference (EuRAD)*. IEEE, 2021.

[58] S. Yan, P. Soh, and G. Vandenbosch, "Wearable Ultrawideband Technology—A Review of Ultrawideband Antennas, Propagation Channels, and Applications in Wireless Body Area Networks," *IEEE Access*, vol. 6, pp. 42 177–42 185, 2018.

[59] Federal Communications Commission, *Part 15*, §15.503. [Online]. Available: https://ecfr.federalregister.gov/current/title-47/chapter-I/subchapter-A/part-1 5/subpart-F/section-15.503

[60] F. Trösch, C. Steiner, T. Zasowski, T. Burger, and A. Wittneben, "Hardware Aware Optimization of an Ultra Low Power UWB Communication System," in *2007 IEEE International Conference on Ultra-Wideband*, 2007, pp. 174–179.

[61] M. Owen. Everything you need to know about Ultra Wideband in the iPhone 12 and HomePod mini. [Online]. Available: https://appleinsider.com/articles/20 /10/18/everything-you-need-to-know-about-ultra-wideband-in-the-iphone-12-a nd-homepod-mini

[62] M. Di Benedetto, T. Kaiser, A. F. Molisch, I. Oppermann, C. Politano, and D. Porcino, *Ultra-wideband Communication Systems: A Comprehensive Overview*. Hindawi, 2006.

[63] T. Zwick, W. Wiesbeck, J. Timmermann, and G. Adamiuk, Eds., *Ultra-wideband RF System Engineering*, ser. EuMA High Frequency Technologies Series. Cambridge University Press, 2013.

[64] "IEEE Standard for Low-Rate Wireless Networks," *IEEE Std 802.15.4-2020 (Revision of IEEE Std 802.15.4-2015)*, pp. 1–800, 2020.

[65] H. Lücken, "Communication and Localization in UWB Sensor Networks," Ph.D. dissertation, ETH Zurich, 2008, Zurich, Switzerland.

[66] A. Rabbachin, "Low complexity UWB receivers with ranging capabilities," Ph.D. dissertation, University of Oulu, 2008, Oulu, Finland.

[67] K. Witrisal, G. Leus, G. J. M. Janssen, M. Pausini, F. Troesch, T. Zasowski, and J. Romme, "Noncoherent ultra-wideband systems," *IEEE Signal Processing Magazine*, vol. 26, no. 4, pp. 48–66, 2009.

[68] M. Z. Win and R. A. Scholtz, "Impulse radio: how it works," *IEEE Communications Letters*, vol. 2, no. 2, pp. 36–38, 1998.

[69] A. A. D'Amico, U. Mengali, and E. Arias-de-Reyna, "Energy-Detection UWB Receivers with Multiple Energy Measurements," *IEEE Transactions on Wireless Communications*, vol. 6, no. 7, pp. 2652–2659, 2007.

[70] M. Z. Win and R. A. Scholtz, "Ultra-wide bandwidth time-hopping spread-spectrum impulse radio for wireless multiple-access communications," *IEEE Transactions on Communications*, vol. 48, no. 4, pp. 679–689, 2000.

[71] S. Won and L. Hanzo, "Initial Synchronisation of Wideband and UWB Direct Sequence Systems: Single- and Multiple-Antenna Aided Solutions," *IEEE Communications Surveys Tutorials*, vol. 14, no. 1, pp. 87–108, 2012.

[72] A. Nasir, S. Durrani, H. Mehrpouyan, S. Blostein, and R. Kennedy, "Timing and carrier synchronization in wireless communication systems: a survey and classification of research in the last 5 years," *EURASIP Journal on Wireless Communications and Networking*, vol. 2016, no. 180, 2016.

[73] M. L. Sichitiu and C. Veerarittiphan, "Simple, accurate time synchronization for wireless sensor networks," in *2003 IEEE Wireless Communications and Networking, 2003. WCNC 2003.*, vol. 2, 2003, pp. 1266–1273.

[74] Y. Yao and X. Dong, "Low-Complexity Timing Synchronization for Decode-and-Forward Cooperative Communication Systems With Multiple Relays," *IEEE Transactions on Vehicular Technology*, vol. 62, no. 6, pp. 2865–2871, 2013.

[75] H. Lücken, T. Zasowski, and A. Wittneben, "Synchronization scheme for low duty cycle UWB impulse radio receiver," in *2008 IEEE International Symposium on Wireless Communication Systems*, 2008, pp. 503–507.

[76] D. Pozar, *Microwave Engineering*, 4th ed. John Wiley & Sons, Ltd, 2012.

[77] C.-C. Lin and H.-R. Chuang, "A 3-12 GHz UWB Planar Triangular Monopole Antenna with Ridged Ground-Plane," *Progress In Electromagnetics Research*, vol. 83, pp. 307–321.

[78] C. A. Balanis, *Modern Antenna Handbook*. John Wiley & Sons, Ltd, 2008.

[79] D.-H. Kwon and Y. Kim, "Suppression of Cable Leakage Current for Edge-Fed Printed Dipole UWB Antennas Using Leakage-Blocking Slots," *IEEE Antennas and Wireless Propagation Letters*, vol. 5, pp. 183–186, 2006.

[80] E. Gueguen, F. Thudor, and P. Chambelin, "A low cost UWB printed dipole antenna with high performance," in *2005 IEEE International Conference on Ultra-Wideband*, 2005, pp. 89–92.

[81] Laird, *Eccosorb LS: Lossy, Flexible, Foam Microwave Absorber Data Sheet.* [Online]. Available: https://www.mouser.ch/datasheet/2/987/RFP-DS-LS_07 0116-1839504.pdf

[82] P. Cao, Y. Huang, J. Zhang, and Y. Lu, "A comparison of planar monopole antennas for UWB applications," in *2011 Loughborough Antennas & Propagation Conference*, 2011, pp. 1–4.

[83] E. Tammam, K. Yoshitomi, A. Allam, M. El-Sayed, R. Pokharel, and K. Yoshida, "Design and analysis of a compact size planar antenna for UWB applications," in *2012 6th European Conference on Antennas and Propagation (EUCAP)*, 2012, pp. 2811–2814.

[84] M. N. Hasan and M. Seo, "A Planar 3.4 -9 GHz UWB Monopole Antenna," in *2018 International Symposium on Antennas and Propagation (ISAP)*, 2018, pp. 1–2.

[85] C. Sulser, "UWB-C Antenna Measurements," 2019, ETH Zurich, internal IKT documentation, unpublished.

[86] Rohde & Schwarz, *ZNBT Vector Network Analyzer Specifications*.

[87] ——, *ZN-Z154 Calibration Unit User Manual*.

[88] X. Lu, X. Chen, G. Sun, D. Jin, N. Ge, and L. Zeng, "UWB-based Wireless Body Area Networks channel modeling and performance evaluation," in *2011 7th International Wireless Communications and Mobile Computing Conference*, 2011, pp. 1929–1934.

[89] Y. P. Zhang and Q. Li, "Performance of UWB Impulse Radio With Planar Monopoles Over On-Human-Body Propagation Channel for Wireless Body Area Networks," *IEEE Transactions on Antennas and Propagation*, vol. 55, no. 10, pp. 2907–2914, 2007.

[90] H. Schulten, "Localization and Posture Detection via Magneto-Inductive and Relay-Aided Sensor Networks," Ph.D. dissertation, ETH Zurich, 2022, Zurich, Switzerland.

[91] J. Bremer, "An algorithm for the rapid numerical evaluation of bessel functions of real orders and arguments," *Advances in Computational Mathematics*, vol. 45, pp. 173–211, 2019.

[92] J.-T. Zhang, "Approximate and Asymptotic Distributions of Chi-Squared: Type Mixtures with Applications," *Journal of the American Statistical Association*, vol. 100, no. 469, pp. 273–285, 2005.

[93] S. L. Brunton and J. N. Kutz, *Data-Driven Science and Engineering: Machine Learning, Dynamical Systems, and Control*. Cambridge University Press, 2019.

[94] F. Pedregosa, G. Varoquaux, A. Gramfort, V. Michel, B. Thirion, O. Grisel, M. Blondel, P. Prettenhofer, R. Weiss, V. Dubourg, J. Vanderplas, A. Passos, D. Cournapeau, M. Brucher, M. Perrot, and E. Duchesnay, "Scikit-learn: Machine Learning in Python," *Journal of Machine Learning Research*, vol. 12, pp. 2825–2830, 2011.

[95] scikit-learn: Machine Learning in Python. [Online]. Available: https://scikit-learn.org/stable/

[96] G. Hinton and S. Roweis, "Stochastic neighbor embedding," in *Proceedings of the 15th International Conference on Neural Information Processing Systems*, ser. NIPS'02. Cambridge, MA, USA: MIT Press, 2002, p. 857–864.

[97] L. Breiman, "Random Forests," *Machine Learning*, vol. 45, pp. 5–32, Oct. 2001.

[98] J. H. Friedman, "Greedy function approximation: A gradient boosting machine." *The Annals of Statistics*, vol. 29, no. 5, pp. 1189–1232, 2001.

[99] H. Sakoe and S. Chiba, "Dynamic programming algorithm optimization for spoken word recognition," *IEEE Transactions on Acoustics, Speech, and Signal Processing*, vol. 26, no. 1, pp. 43–49, 1978.

[100] D. J. Berndt and J. Clifford, "Using Dynamic Time Warping to Find Patterns in Time Series," in *Proceedings of the 3rd International Conference on Knowledge Discovery and Data Mining*. AAAI Press, 1994, p. 359–370.

[101] S. Salvador and P. Chan, "Toward Accurate Dynamic Time Warping in Linear Time and Space," *Intell. Data Anal.*, vol. 11, no. 5, p. 561–580, Oct 2007.

[102] F. Petitjean, G. Forestier, G. I. Webb, A. E. Nicholson, Y. Chen, and E. Keogh, "Dynamic Time Warping Averaging of Time Series Allows Faster and More Accurate Classification," in *2014 IEEE International Conference on Data Mining*, 2014, pp. 470–479.

[103] A. Mueen, N. Chavoshi, N. Abu-El-Rub, H. Hamooni, A. Minnich, and J. MacCarthy, "Speeding up dynamic time warping distance for sparse time series data," *Knowledge and Information Systems*, vol. 54, p. 237–263.

[104] Z. Geler, V. Kurbalija, M. Ivanović, M. Radovanović, and W. Dai, "Dynamic Time Warping: Itakura vs Sakoe-Chiba," in *2019 IEEE International Symposium on INnovations in Intelligent SysTems and Applications (INISTA)*, 2019, pp. 1–6.

[105] V. Mahato, M. O'Reilly, and P. Cunningham, "A Comparison of k-NN Methods for Time Series Classification and Regression," in *AICS*, 2018.

[106] wannesm, khendrickx, A. Yurtman, P. Robberechts, D. Vohl, E. Ma, G. Verbruggen, M. Rossi, M. Shaikh, M. Yasirroni, Todd, W. Zieliński, T. Van Craenendonck, and S. Wu, "wannesm/dtaidistance: v2.3.5," Jan. 2022. [Online]. Available: https://doi.org/10.5281/zenodo.5901139

[107] R. Tavenard, J. Faouzi, G. Vandewiele, F. Divo, G. Androz, C. Holtz, M. Payne, R. Yurchak, M. Rußwurm, K. Kolar, and E. Woods, "Tslearn, A Machine Learning Toolkit for Time Series Data," *Journal of Machine Learning Research*, vol. 21, no. 118, pp. 1–6, 2020.

[108] "IEEE Standard for Low-Rate Wireless Networks–Amendment 1: Enhanced Ultra Wideband (UWB) Physical Layers (PHYs) and Associated Ranging Techniques," *IEEE Std 802.15.4z-2020 (Amendment to IEEE Std 802.15.4-2020)*, pp. 1–174, 2020.

[109] K. Kira and L. Rendell, "A Practical Approach to Feature Selection," in *Machine Learning Proceedings 1992*, D. Sleeman and P. Edwards, Eds.   San Francisco (CA): Morgan Kaufmann, 1992, pp. 249–256.

[110] C. Ding and H. Peng, "Minimum Redundancy Feature Selection From Microarray Gene Expression Data," *Journal of Bioinformatics and Computational Biology*, vol. 03, no. 02, pp. 185–205, 2005.

[111] X. He, D. Cai, and P. Niyogi, "Laplacian score for feature selection," in *Advances in Neural Information Processing Systems*, Y. Weiss and B. Schölkopf and J. Platt, Ed., vol. 18.   MIT Press, 2006.

[112] W. Yang, K. Wang, and W. Zuo, "Neighborhood Component Feature Selection for High-Dimensional Data," vol. 7, no. 1, pp. 161–168, 2012.

[113] R. Kohavi and G. John, "Wrappers for feature subset selection," *Artificial Intelligence*, vol. 97, no. 1, pp. 273–324, 1997.

[114] C. Shorten and T. Khoshgoftaar, "A survey on Image Data Augmentation for Deep Learning," *Journal of Big Data*, vol. 6, 2019.

Bisher erschienene Bände der Reihe
Series in Wireless Communications

ISSN 1611-2970

Alle erschienenen Bücher können unter der angegebenen ISBN-Nummer direkt online
(http://www.logos-verlag.de) oder per Fax (030 - 42 85 10 92) beim Logos Verlag
Berlin bestellt werden.